PRAISE FOR *EAVESDROPPING ON ANIMALS*

"An engaging guide to a world of wonders hiding in plain sight."

PETER WOHLLEBEN, *New York Times* bestselling author of
The Inner Life of Animals and *The Hidden Life of Trees*

"In this warm and welcoming book, Bumann engages all
our senses to help bring us back home again, reconnecting us
to our fellow creatures on this sweet, green Earth. You won't
want to miss the conversations going on around you!"

SY MONTGOMERY, author of *Of Time and Turtles*
and *The Soul of an Octopus*

"This lovely friendly book is your personal invitation
to what Bumann calls 'an inescapable sense of awe.'"

CARL SAFINA, author of *Alfie and Me:*
What Owls Know, What Humans Believe

"An enjoyable and timely reminder that
there is another world we once heard and might
yet hear again—if we take the time to listen."

TRISTAN GOOLEY, author of *How to Read a Tree*

"Jaunty, savvy, learned, and compulsively
readable. Bumann teaches us how to pay attention—
not just to the wild world, but also to ourselves."

CHARLES FOSTER, author of *Being a Beast* and *Cry of the Wild*

"This gracefully written, captivating book
guides us in keeping our senses wide open when we
venture into nature. Despite my fifty years' experience
as a naturalist, I gained much new awareness, and I am
already putting Bumann's teachings to good use."

JONATHAN BALCOMBE, author of *Super Fly*
and *What a Fish Knows*

"There are dozens of books that tell us how to watch wild animals, but only one that explains how to listen to them. In *Eavesdropping on Animals*, Bumann shows us how to unlock the code of animal communication and become privy to the fascinating ways animals talk to their friends, their foes... and to us. A wonderful read!"

DOUGLAS W. TALLAMY, author of *Nature's Best Hope*

"Rewilding our hearts and souls in the ways laid out in this book will go a long way toward improving people's lives as well as the health and well-being of our magnificent but wounded planet."

MARC BEKOFF, PhD, author of *Rewilding Our Hearts: Building Pathways of Compassion and Coexistence* and *The Emotional Lives of Animals: A Leading Scientist Explores Animal Joy, Sorrow, and Empathy—and Why They Matter*

"Learning to eavesdrop from Bumann, we improve our chances of sighting rare wildlife and come to realize that our presence and actions are anything but secret to our wild neighbors."

JOHN M. MARZLUFF, professor emeritus of wildlife science, University of Washington, and author of *Gifts of the Crow, Welcome to Subirdia*, and *In Search of Meadowlarks*

"There is no doubt in my mind that reading Bumann's book will enlarge your ability to appreciate nature and add to your enjoyment of life in general."

ROBERT BATEMAN, Canadian naturalist and painter

"Bumann combines his biologist's training with an artist's sensitivity to describe strategies we can apply to engage, understand, and more fully appreciate the many different ways wild species communicate. His evidence, insights, and guidance encourage us to apply all our senses—and suddenly we move beyond mere identification and into the complexity of life."

TONY ANGELL, author of *The House of Owls*

"In *Eavesdropping on Animals*, naturalist and artist George Bumann has provided a vital off-road map for all of us to find our way—and even perhaps our place—in the natural world."

JOE HUTTO, award-winning author of *Illumination in the Flatwoods*

"I've spent my entire life listening to the natural world, and after reading Bumann's book, I realize I wasn't really listening at all."

JACK HORNER, presidential fellow, Chapman University

"This book left me yearning to be in a forest, leaning into the conversations around me."

JOANNA E. LAMBERT, animal behavioral ecologist and professor, University of Colorado Boulder

"Combining his personal experiences listening to animals with scientific discoveries about how they communicate, Bumann asks—in the tradition of any good naturalist—what can we hear, what can we see, what can we experience? The result is a highly readable book that clearly demonstrates there is a whole world of animal language all around us, waiting for us to figure out how to listen and how to understand."

CON SLOBODCHIKOFF, PhD, author of *Chasing Doctor Dolittle: Learning the Language of Animals*

"Bumann's book is not only an enlightening exploration of animal behavior but also a celebration of the magic inherent in our shared existence with countless curious beings."

KATIE SIEVING, professor of wildlife ecology and conservation, University of Florida

"In *Eavesdropping on Animals*, Bumann gives you the tools and skills you need to build a closer relationship with the natural world. And more importantly, this elegant volume will make you excited to go outside and open your new toolbox. It will improve your life."

SCOTT MCMILLION, author of *Mark of the Grizzly* and editor of the *Montana Quarterly*

GEORGE BUMANN

Foreword by JON YOUNG

EAVESDROPPING ON ANIMALS

What We Can Learn From Wildlife Conversations

GREYSTONE BOOKS

Vancouver/Berkeley/London

Greystone Books Ltd.
greystonebooks.com

Cataloguing data available from Library and Archives Canada
ISBN (cloth) 978-1-77840-020-9
ISBN (epub) 978-1-77840-021-6

Editing by Jane Billinghurst
Copyediting by Brian Lynch
Proofreading by Jennifer Stewart
Jacket and text design by Fiona Siu
Jacket photograph by GarysFRP/iStock.com

Printed and bound in Canada on FSC® certified paper at Friesens.
The FSC® label means that materials used for the
product have been responsibly sourced.

Greystone Books thanks the Canada Council for the Arts,
the British Columbia Arts Council, the Province of British Columbia
through the Book Publishing Tax Credit, and the Government of
Canada for supporting our publishing activities.

Canadä

MIX
Paper | Supporting
responsible forestry
FSC® C016245

BRITISH COLUMBIA

BRITISH COLUMBIA
ARTS COUNCIL
An agency of the Province of British Columbia

Canada Council   Conseil des arts
for the Arts   du Canada

Greystone Books gratefully acknowledges the xʷməθkʷəy̓əm (Musqueam),
Sḵwx̱wú7mesh (Squamish), and səlilwətaɬ (Tsleil-Waututh) peoples on
whose land our Vancouver head office is located.

To my grandfather—you were my hero

For my mother and in memory of my father—
you made so much possible

And to Jenny and George—
you are a constant source of joy and
inspiration. It never ceases to amaze me
how much you show me of this world

CONTENTS

FOREWORD

OMBINING AGE-OLD PRACTICES and contemporary science with personal observations, George Bumann places the art of listening to animals and interpreting what they have to say firmly within anyone's grasp.

George's talent and commitment to learning and communicating emerged for me when he attended a bird language workshop in coastal California. My role was to draw participants out to offer their observations. It quickly became evident that George belonged with the instructors, and he has been part of our global bird and animal language teaching team ever since. Since then, George and I have collaborated at every chance we could find, and we often share recordings about animal language to get each other's reflections and questions for further inquiry.

When I published *What the Robin Knows: How Birds Reveal the Secrets of the Natural World* back in 2012, with the help of science editor Dan Gardoqui, there was relatively little science to back up our observations. In the ensuing decade, a flourishing of research emerged on interspecies communication. In this book, George has picked up where we left off, combining the kinds of observational practices I learned from my mentor, a master tracker, with cutting-edge scientific research into bioacoustics, the science of how animals communicate.

George is never afraid to share his theories with experts to harden his (and our) understanding of the principles he is investigating. Thanks to his ever-expanding collaborations with holders of traditional knowledge, his distillation of scientific research, and his field experience and direct observation, he continues to grow the edge of our understanding of the language of birds and animals. Taking this one step further, George is now dedicated to sharing his knowledge about ways in which understanding animal communication can benefit everyone from members of the military on active duty, to intrepid travelers, to individuals wanting to connect more closely with the places they call home.

The results of what George has researched, learned, and experienced are encapsulated in this highly readable book. *Eavesdropping on Animals* provides readers with an entry point into decoding the nuances of nonhuman communication, relating this not only to birds—including those with an especially challenging range of vocabulary, such as the ravens he encounters almost daily in Yellowstone—but also to a wide range of animals both exotic and commonplace. With the profound wisdom born of experience, George leads us on a journey that builds slowly and draws us across a threshold from being strangers looking in from the outside to becoming members of the family of living beings we share this world with—truly a life-changing transformation. After absorbing the lessons in this book, you will never feel the same way about the natural world, even in your own backyard.

JON YOUNG

INTRODUCTION

A WOLF IS OUT THERE. Although I cannot see it or hear it, the wild canine is definitely close. The questions are precisely what wolf and exactly where. With each passing moment, updates of its presence pour in. *Canis lupus* is not at the doorstep, but no more than two or three hundred yards to the east. Insights continue to arrive. As the seconds lead to minutes, it becomes clear that the wolf, or maybe even a pack of wolves, is just over the hill beyond Yellowstone Park's helicopter landing pad—and it is on the move.

It is December 11 and the frozen expanse of the world's first national park stretches out around me. Under clouded, wintry skies, the earth is a collage of white snow and sagebrush gray. I still have not spotted the wolf, nor do I have the benefit of intel from tracking devices or information from other people. Is this the Yellowstone Canyon Wolf Pack on its first foray into the park's northern range this year, I wonder? And if so, how many wolves are there? The questions multiply. By now, the time is 8:21 AM. What I am sure of, standing alone with my senses, is that the wolf, or wolf pack, is traveling from right to left at a pace of roughly four to six miles per hour and heading downstream along the Gardner River. How do I know?

A single vociferous coyote aired the breaking news about the wolf. I could easily have tuned out its warning bark and dismissed it as senseless background noise. Most of us spend our entire lives turning a blind eye and deaf ear toward animal conversations, chalking the chatter up to just more meaningless racket. The result of not letting the wild dialogues in is that we have been missing out.

Entire communities of organisms have been communicating in complex ways for millions of years. Decoding the details is akin to tapping into the most ancient social media, the first Facebook, the original Twitter. This knowledge network is transmitting information twenty-four hours a day, 365 days a year, at speeds of one hundred miles per hour or more. Wildlife conversations are occurring across the urban-wilderness divide, and bridge species, genus, family, order, and even class boundaries of organisms. The warblers are listening to the squirrels; the squirrels are listening to the crows. The crows are lending an ear to the crickets and the deer—and they all may be hanging on every peep the frogs have to say. And there is more. Amid the back and forth, there are those who have the community at heart, but maintain personal agendas; there are mind readers and prognosticators, broadcasters of five-alarm warnings and silent observers. Everyone in the neighborhood is part of this collective vibe—everyone, that is, except us.

If the average person is aware of the voices of nonhumans, the conversations they are most likely to notice are the calls collectively known as the dawn chorus—that loud, seemingly disorganized ensemble of trilling, whistling, crowing, chirping, honking, and shrieking that seems to do everything but hush up when you're trying to get some sleep. Many a window has been thrown open, much verbal effluent spewed, and numerous projectiles thrown at some unsuspecting bird just trying to get a date. Some people resort to earplugs or a well-placed pillow, while others retreat into their personal auditory oasis, as a friend of mine does in his luxury car, which monitors ambient sound levels and increases the volume on the radio to cover up outside noise. We go to great lengths to control what enters our ears and our minds, and as we do so, we seldom regard animal noise for what it truly is—communication.

Later in the evening, I post on social media that the Canyon Wolf Pack might be in town. A friend and safari guide replies that he was out with some clients that morning and spotted the pack near the bridge along the Gardner River, just east of Mammoth Hot Springs—two miles upstream from my location. They lost sight of them around 8 AM, moving downstream in my direction.

Although I was happy for the confirmation, I didn't really need this third-party report to know that my suspicions were on target. Everything in this morning's encounter had been writ large across the landscape. The notifications were not pinged through in a cell-phone message or drafted in ink; rather, they were transmitted in sound. Had you been standing next to me and identifying the relevant clues, you too would have known the large predator was near.

At this point, you might wonder if I am overstating my case. How could I be so certain of these details of species, distance, and direction? It sounds a bit far-fetched, doesn't it? Can human bystanders really slip into the shoes of Doctor Dolittle or access the equivalent of living, breathing surveillance cameras for a front-row seat to hidden dramas? As it turns out, we most certainly can. The pivotal clue that morning, the auditory signal that broadcast the pack's presence, was the coyote's unique barking howl, a call that is reserved exclusively for situations like this—it is the coyote word for "wolf."

You can perform similar acts of wildlife spotting in your own community, along most any street, woodlot, or town park. Once you learn how to tune in to the wild conversations around you, though the animals might be different, the parallels between what you discover as you relax on your patio or gaze out the window sipping your morning coffee and what you can see in distant locations such as Yellowstone or the African savanna will be uncanny. In fact, some of the best preparation for a once-in-a-lifetime trip to a far-off destination can be done within a block of home by relying on nothing but your five senses.

Although our ability to explore the world's biodiversity has taken a giant leap forward in recent times—thanks to high-quality field guides, advanced optics and recording equipment, reference databases, and artificial intelligence—you don't need any special aids to immerse yourself in nature. Even when you have left your devices behind or their batteries are running low, you can still participate in the events going on around you, perhaps even more effectively than through the filter of technology. Even for those of us whose sensory abilities have become a bit rusty with age or atrophied from disuse, our innate potential to tune in is still nothing short of amazing, and people with perceived shortcomings such as limited mobility or diminished hearing or eyesight can experience the outdoors in ways that leave others in the dust.

As you start to study your undomesticated neighbors, you will soon realize that they are also eyeing you. Animals are sharing many tidbits about *you* personally, distinct from your friends, neighbors, and family. Think of wild conversations as an alternate form of neighborhood watch or as primordial big data. Although animals may not be logging your internet activity, they are taking in far more than you realize. As your awareness begins to shift, you begin to get a sense of how nonhuman others might perceive the world and how they perceive you—and your place within it begins to change.

When the human and nonhuman spheres overlap, exciting things happen. Meeting individual nonhumans—such as the crow with a discolored tail feather that hangs out by the park bench or the female deer with the nick in her ear that comes to your yard each evening—is how some of the most transformative experiences arise. With time, you will appreciate how these seemingly distant others are much more like us than not. See how fiercely they defend their young or how tenderly they sweet-talk their mates. Discover for yourself how local creatures rely on their communities, honor clannish traditions, depend on certain habitat features, and retrace ancestral pathways.

Learning how to listen has added immeasurably to my own life as a parent, homeowner, artist, educator, and curious observer. I invite you to join me for a journey through concrete jungles, suburban grasslands, and distant corners of the planet's wildscapes. Not only will the insights and lessons thrill and surprise you, but you will never look at your own yard, city, or neighborhood park the same way again. The wilds are full of wisdom and intrigue, and you may find that you become equal parts detective, poet, and eager pupil amid the timeless relationships of the outside world.

— PART I —

A SENSE FOR YOUR SURROUNDINGS

— 1 —

VOICES
ALL AROUND

PHILIP WAS STUNNED and a little disoriented by what he was hearing. Silence, or an eerie version of it, had replaced the usual cacophony of the city. Birdsong echoed around him in a way he had never heard before. Wind gently whispered through the trees. Although the quiet was a tad unsettling and the identity of the birds was lost on him, Philip knew this was special. Amazing, actually. He was not standing in a woodland preserve; Philip was seated at his desk next to the open window of his home office in Brooklyn, New York. This was about as far from wilderness as one could imagine. And yet, even though he was in an urban residential area not far from the heart of the Big Apple with its maze of concrete, glass, and steel, here they were, multitudes of creatures belting out their arias as if the concert hall was entirely theirs. And it was.

A publishing editor by trade, Philip had been relegated, like many, to cobbling together a workspace in his cramped studio apartment. It was March 2020, the coronavirus had just reached American shores, and the entire city was on lockdown. Scarcely a peep of pedestrian or automobile traffic could be heard, and there was an unsettling sense that the apocalypse had arrived and settled on the city. Without the perennial office soundtrack of shuffling papers, tapping keyboards, and secondhand phone conversations, Philip discovered something new, something usually buried within the din of metropolitan living. Others noticed it too, and some remembered the echo of a similar quiet in the aftermath of the 9/11 attacks on the World Trade Center. Philip paused. He pushed back from his desk, closed his eyes, and listened.

With regular schedules thrown on their ear, many of us, like Philip, were discovering what had been hiding in plain sight all along. New York was not the only place it was happening. Around the globe, while people were binge-watching Netflix series and newsfeeds of COVID coverage, telecommuting in sweatpants with disheveled hair while tied to caffeine IV drips as kids pounded on home office doors, and attending Zoom calls with little or nothing on below the waist, life carried on—outside.

With unrelenting congestion, mind-numbing traffic, and constant human activity on hold—in what some would dub the Anthropause—animals lived their lives in ways seldom seen by most people. Wild boars were entering the inner sanctum of Rome and leopards were seeking easy prey in the outskirts of Bengaluru, the megacity formerly known as Bangalore, India. The lack of noise pollution in the San Francisco Bay Area allowed white-crowned sparrows to literally sing a different tune. Instead of shouting over busy street vendors, honking taxis, delivery trucks, and construction noise to delineate their territories and attract mates, the birds were able to sing a richer, more delicate and beautiful brand of song. And back home in Yellowstone, while my family and I were on Long Island, New York, our ranger friends were reporting that wolves were making kills right next to the roads, grizzly bears were sleeping in parking areas, and bison were bedding down by the hundreds across the pavement at all times of the day and night.

Imagine what you might hear or see the next time the hush returns. Honestly, don't wait for that day to come. Try sifting through the layers of what is around you right here, right now. When was the last time you stepped outside and just listened? Picture yourself pausing just long enough to hear or see something unexpected at a location close to home. You might have to do this on a weekend or after the rush-hour traffic has ebbed, but try it. Who knows, you might detect a frog croaking or a squirrel chattering where you have never noticed one before.

Opening our senses to a wider world can both puzzle and excite us. Although large, garish, or loud creatures often dominate our attention—such as the bellowing alligator in a golf course pond or the caterwauling macaw in the jungles of Costa Rica—make a point of not overlooking the softer, subtler speakers. Every voice in the choir plays a part and many *aha* moments are going to be triggered by the diminutive, the drab, the meek, and the unseen. Consider that the frog or squirrel you just heard could be revealing the presence of a snake, an owl, or some other, even more exciting and unexpected resident.

Let it all in. Even if your soundscape includes human-made noise, this will give you an appreciation of the conditions under which our wild neighbors live. They can't turn the cacophony off, and hence their lives literally depend on listening to the full range of sounds surrounding them. Alarms announcing the approach of a lethal predator cannot be disentangled from the noise of a diesel truck—and the predator might be using the sound of the truck to mask news of its arrival, thus allowing it to get much closer to its prey.

Paying attention cultivates greater awareness of everything the soundscape offers, with or without those forced moments of pause that periodically land upon our doorsteps. As for the specific details of what animals are saying and how to interpret those conversations, we will get to those. For now, take a moment to immerse yourself in the moment as Philip did, and feel the thrill of listening.

— 2 —

ENTERING CREATURE-SPEAK

THE FIRST EXPERIENCE I had entering creature-speak came when I was quite young. On one of my first overnight expeditions into our mowed half-acre lawn in upstate New York, I awoke in the morning to blankets glittering with dew and an unexpected sound. I knew very few birds at that time, but one I did know—the American robin, a group of them actually—was in an uproar. The robins around our home usually seemed content bouncing around the cropped grass and engaging in tugs-of-war with worms. These birds, by contrast, were jumping around and chirping in a mad fury over something I could not see.

I was vaguely aware that there was a robin's nest in the neighbor's disheveled willow tree, so I kept watching. Then I saw it. Something black appeared. Whatever it was, it was *in* the nest. The object grew, inflating like a dark, iridescent balloon until it took flight in the form of a crow. The crow held within its beak a robin nestling. As the thief turned in midair, I could see the little citrus-orange feet of the young bird kicking as they disappeared.

A few shrieking robins followed, but there was nothing left to do. It was over.

When the action dissipated, I found that I had been clenching my blankets and holding my breath the entire time. The yard felt eerily empty afterward, diminished in some way. For the robin parents, their lives had been shattered. For me, this was a turning point. The whole episode, which lasted all of thirty seconds, sticks with me as vividly now, four decades later, as it did in the immediate aftermath. I had no idea what those birds were saying, but I remember feeling a distinct sense of dread. These wild creatures, ones I had scarcely noticed, had lives of their own. Their days came complete with fears, needs, memories, and motives that were separate from, but parallel to, mine.

As much of a game changer as that encounter was for me, I felt a sort of malaise during the ensuing decades. I grew up in the countryside in a culture of hunting and fishing, and there was a limited number of options for what one did and did not do outdoors. The idea of going for a walk with the singular purpose of listening to birds, for example, was unheard-of. Within the known possibilities, I did whatever I could in search of more of what I found in the willow that morning. I fished religiously and ran traplines. I hunted with guns and bows with a fervor and degree of success that many hunters would envy. In short, I learned to become exceedingly lethal. Yet I remained distanced from the essence of something I could feel but not quite define.

To be clear, I benefited greatly from all my time in the woods. There were countless lessons I could learn only there. I absorbed essential tenets of conduct such as *placing* my feet as I walked instead of simply stepping (to avoid breaking sticks and kicking stones) or looking around from *inside* the forest edge before stepping into the open. When I moved, I learned to use the terrain to conceal my advance—traveling across exposed ridges or open fields can cause a group of deer, turkeys, or elk to sound the alarm from a half a mile away or more. And I always checked the wind, not once, but multiple times. I don't think there's ever been a time when the wind ruled my life more than when I was a hunter. Think of these precepts like rules of etiquette. Imagine attending an upscale dinner and picking up the wrong fork; the break in protocol creates tension and results in missed opportunities.

Despite "becoming animal" in so many ways in order to pursue them, it would be a lie to say my cup was full. I yearned for deeper understanding. Many of my relatives and role models did not go into the woods outside hunting season, which only added to my sense of disillusionment. Did they not have their own robin experiences? I was forced to reflect on my role as the grim reaper; so much of it seemed like a colossal missed opportunity. No other paths appeared on the horizon and there were no teachers.

What was clear was that I found myself inextricably drawn to wild animals that I could have a conversation with. In a time before YouTube became synonymous with how-to, I sought out the best of the best animal impersonators: the reigning New York state turkey-calling champion; a friend of my grandfather who was a top-tier, competition-tested voice double for ducks; a family acquaintance who taught me the subtleties of using a goose call the way a band teacher might instruct a pupil on clarinet or saxophone. Birds were not my only target either; deer and other animals such as foxes and raccoons did not escape my imitative impulses. I even figured out a way to call North America's largest rodent, the beaver. Beyond using my own vocal cords, I built devices out of wood, bone, plastic, stone, and rubber with which to expand my repertoire to more than two hundred species of living things. I was obsessed. It all eventually came to a head in the state capital, Albany.

I was twenty-one and had never driven to this part of the state before. I feared missing the New York State Thruway exit as misting rain turned into a downpour. I was on the road to attend an event being held in a dark, low-ceilinged hotel convention center with an uninspiring brown-on-brown motif. The trip took me halfway across the state to conduct the serious business of making turkey sounds for a couple of men hiding behind a curtain. This was the National Wild Turkey Federation's sanctioned New York State Turkey Calling Contest—yes, such a thing exists.

Before it was my turn to perform, I toured the various banquet rooms, sifting through crowds of people clad in camouflage and weaving my way between tables laden with turkey calls and other hunting paraphernalia. When my number was called, I took the stage and assumed the appropriate spot in front of the crowd. The emcee leaned into his microphone and in a dry monotone voice prompted my performance by saying, "Yelp of a hen turkey."

The judges were, I presumed, a couple of men with score cards behind a curtain stage right. I shifted the call in my mouth—a device roughly the size of a sacrament wafer made of athletic tape with a bit of latex, and looking something like a mix between what you'd use to cover a heel blister and a contraceptive device. I rendered my best yelp as I had rehearsed it for months. It was then followed by my best "tree call," "excited cutting of a hen," "assembly call," "fly down cackle," "kee kee run," "cluck and purr," and, of course, a gobble. Should you cringe at the thought of your child practicing trumpet in the house, or cover your ears amid the pounding of drums, the screeching of violins, or—god forbid—the keening of bagpipes, these sounds don't hold a candle to what comes out of an aspiring turkey caller.

I managed to head home with a trophy topped with a gilded plastic gobbler for a second-place finish in the state's amateur open division. I also took home one of the top honors in the natural voice section—a subset of the contest where no calling devices are allowed save your own vocal cords. I came away from that experience with something much more memorable than a trophy. It was a notion of how ridiculous these contests are.

Number one, actual turkeys, to my knowledge, have never been invited to jury said calling events. Secondly, and by the admission of organizers and most participants, real turkeys would probably never win. The absence of actual turkeys and their perspectives was a conspicuous omission to me. What were the turkeys themselves doing at this very moment? The fact was, I didn't know. This event built upon a framework of what human beings believed turkeys did and did not do, with the aim of putting one on the dinner table. Some years later, I would find a *Washington Post* article interviewing one of the judges for the Maryland state calling championship. In it he stated, "If you hear really good calling for a long time, you can almost be sure it's a hunter, not a turkey. A turkey makes a lot of mistakes."

Wait, hold on a moment. Mistakes? Would a turkey not say exactly what it meant to say? Isn't this a bit like making a trip to a fast-food joint, placing an order for a burger and fries, and then having some guy randomly leap out from behind a curtain to inform you, "No, no! You didn't mean to say a burger and fries.

You meant to order a toothbrush and a sewing machine"? Say what? This would not only be my first turkey calling contest, it would also be my last. And with that, I retired.

I had entered into "creature speak" enough with my own vocal cords and these devices to have a notion of what these animals say, but I felt as far from understanding the world of another species as ever. I sensed I was approaching the whole thing from the wrong direction. The contest was an opportunity for me to compete with other human turkey impersonators; it did not help me understand the birds. What I really wanted to do was probe the depths of the turkeys' world as they themselves might experience it. I wanted to know what really makes them tick, not what the judges deemed to be a perfect turkey call.

Out of my ruminations came other realizations, some of which I was loath to admit. Among them was the sense that the challenge of bagging one's quarry does not require monumental insight. Aside from modern hunting involving long-range firearms and military-grade technology, once the trigger is pulled, the learning stops. This fact was not lost on the twenty-sixth president of the United States and conservation icon Theodore Roosevelt either. In his travels, TR interviewed many of the venerated hunters of his day. The conclusion that Roosevelt came away with was that most hunters were "very untrustworthy in dealing with points of natural history... [and] only know so much about any given animal as will enable them to kill it."

This is not a dig against hunters, as they are some of the most dedicated conservationists; their direct actions and funding have protected and enhanced large amounts of wildlife habitat. Rather, it illustrates the fact that this approach is but one lens through which to view nature. For myself, going afield during hunting season was not enough. My thoughts kept wandering back to the birds, the beavers, and the deer. What might they think?

My longing was for something bigger, more comprehensive. I wanted something divested of human artifice and rooted in an unfiltered exchange with my surroundings and those other sensing, feeling beings. As the author David Brooks said, "The point is not just to describe the fish in the river, but the nature of the water in which they swim." Diving deeper into animal conversations would prove to be the on-ramp I was looking for. My experiences sharing this approach in ensuing decades have only reinforced my conviction that this approach works for others as well.

So how do you enter the illuminating world of animal conversations? It's all about rekindling a sense of childlike curiosity and then finding spaces where you can nourish it. First, identify the two or three places where you spend most of your time—even if they are indoors. Then ask yourself how many of the routine activities you perform in those places can be moved near a door or window or, better yet, outside. Can you hold regular phone conversations in the courtyard, as a friend of mine does? Or make an effort to dine in the open air for at least one meal every day?

What might you automatically be aware of if your windows were open? If cold weather or biting insects are an issue, try what a friend of mine does and set up a wireless microphone system like those used at nature centers to pipe outdoor sounds in. If your interests and curiosity compel you, go to a local park—just make sure whatever outside space you choose is easily accessible, otherwise you won't go. More dedicated practices, such as a "sit spot" routine, will help expand and deepen your experiences, but we will discuss those in due course.

Look at any and all outside time as an opportunity to both actively and passively absorb what is going on around you. Of particular importance is a notion of what constitutes normal. Do you typically see squirrels or are there usually only chipmunks? Are pigeons and starlings common to your local greenbelt or are they only periodic visitors? When was the last time you saw a gecko? Discovering some of the most common creatures can be a revelation in itself. The novelty of finding wild ones is not so much that they are unusual, it's that you are now seeing them for the first time.

Let it all in, have fun, and be curious. Rather than playing the name game with every plant, animal, or fungus, place a premium on noticing tendencies. The good stuff starts when you allow natural rhythms into your everyday. Will you make mistakes along this journey? Absolutely. Will you botch interpretations of animal signals? Again, yes, yes, and yes, but what that means is that you are beginning to pay better attention. You may find yourself confused by what seem like random animal noises, but know this one simple fact: animal sounds and behaviors are tied to the essentials of life, so pay attention, they have meaning. The revelations come when you begin to decipher the signal behind the noise.

— 3 —

HEARING
A PIN DROP

B EFORE WE GO ANY FURTHER, let's take a look at how we explore the world around us. I grew up in my mother's sculpture studio in New York. I followed in her footsteps and am now a sculptor in my own right. When I am not out on the trail with my wife and son, Jenny and George, or walking our Labrador retriever, Hobbes, I spend a lot of time observing wildlife for my work. I invoke the field of art because, like the sciences, it is a powerful approach in the search for truth. An inspiring example of how we can use our senses to grasp these larger truths comes from a sculptor acquaintance of mine.

Floyd DeWitt was twenty-six when he was handed a pin by his professor. The pin was an ordinary thing, average-sized as pins go, the sort you might use for an everyday sewing project. Floyd was a first-year student at the venerable four-hundred-year-old Royal Academy of Fine Art, the Rijksakademie van Beeldende Kunsten, in Amsterdam, Holland. Following his studies at the Minneapolis School of Art, this ranch kid from the big-sky country of eastern Montana was adapting to life and art in Europe in the 1960s.

"Look at this pin. Examine it," said his instructor. To Floyd's eyes, there was not much to note: it was pointed at one end and had a head at the other, silver. "Now give it back." As soon as Floyd complied, the professor drew back his arm and threw the pin over Floyd's shoulder and into the deep shag carpet behind him. He then said, "Go find it."

"Go find it?"

Floyd left his chair and lowered himself to the floor. He searched in vain for many minutes and found what one might expect: nothing. Saving the student from further indignity, the professor said, "Take your seat." Once back in his chair, Floyd watched as the older gentleman got up from behind his desk, walked directly over, and picked up the pin. "Here," he said, handing the pin back to the student.

Floyd was dumbstruck, but not for the reason you might think. This lesson stayed with Floyd for the next sixty years because this professor had been blind since birth. A blind teacher may not be unusual for an academy of music, but for him to be tenured at an institution of visual art seemed strange. The professor brought the matter full circle after resuming his own seat. "You have eyes, but you don't see. You have ears, but you do not hear," he said.

The message resonated like a bell. The point was to impress upon the student that to discover the beauty in life and to create good art, good anything, you have to start by using your senses to their highest potential. Take a detailed inventory, then put them to use. The blind professor went on to explain that during World War II, many watchtowers in Britain enlisted the help of blind listeners. Although most soldiers could hear the German aircraft coming, it was the blind who could best tell what direction the

Luftwaffe was advancing from. On average, the hearing of the blind is not that different from the general population's, but it is unique in that it has been honed through extensive focused use. Floyd would drill the primacy of close study and a refinement of the senses into me over many learning sessions at his studio. What we find with our enriched sensory appetite, Floyd would say, "is an inescapable sense of awe."

When compared with perceptual feats elsewhere in the animal kingdom, human capabilities may seem stunted, but this is too harsh a judgment. Sure, our eyes may never see light in the ultraviolet bandwidth like a bee does, and our ears may not detect ultrasound the way a bat might. Each person, however, is endowed with their own precision-built monitoring systems that can be put to use for all sorts of discoveries.

"Study from nature" was what so many masters of the past preached. Ancient Greek artists, as well as those of the baroque and Renaissance periods, the impressionists, and yes, even the abstract expressionists, studied their environment in great detail. Indicators of careful observation ripple throughout the work of past masters with a vitality that is as palpable today as it was decades or centuries ago. Art and science are much more aligned in this way than they are usually given credit for, and both can help us understand animals. It all comes down to better observation.

The oldest known art depicting animals created by *Homo sapiens*—the warty pigs from Sulawesi, Indonesia, dating to at least 45,500 years ago, and later works in European caves such as Chauvet and Lascaux—show an extraordinary facility for observation. An understanding of anatomy, composition, the depiction of form, foreshortening, and motion existed long before these terms were invented. The renderings on the walls of the Chauvet Cave in France from around 32,000 years ago possessed such veracity that they clarified a mystery for contemporary paleontologists: extinct cave lions, unlike their present-day African counterparts, did not have manes. Modern science can be informed by ancient art.

Those early artists created descriptive and expressive art purely by schooling themselves in the vibrant details of their everyday surroundings. As a matter of survival, early peoples used their senses to the highest degree—and they inspire us to do more of the same now. Art, photography, and journaling are just a few of the activities that might help you focus your attention on things that are otherwise missed.

You don't need to be a skilled photographer, artist, or writer for these activities to work for you. A participant in one of my classes related an exercise that completely changed her view of the forest near her home. As part of a nature journaling class, she was asked to find a natural object and draw it. She chose a Douglas fir cone. The forest around her is full of Douglas firs and she has a number of these iconic trees growing in her backyard. She would say that she is no artist, but she dutifully drew her cone; this was a private exercise just for her, after all. To her amazement, she noticed structures extending from the cone's scales in the shape of the hind legs and tails of mice. These bracts are distinctive identifiers for Douglas fir cones, and she had been cleaning up the cones from her yard for almost a decade and yet had never seen these curiously shaped bracts before. Now she sees them everywhere and can identify Douglas fir cones on sight. One simple act of taking the time to look closely enough to draw them forever fixed them in her memory.

Another great way to begin reclaiming our talents for detection is becoming aware of the special qualities of our peripheral vision. When we gaze upon a world that looks sharp and clear, we are making this judgment based on the central 2 percent of our visual field, from the approximately 180 degrees at our disposal. Our fragment of clarity amounts to an area a little bigger than a thumbnail held at arm's length.

I fondly recall a visit with Dr. D. Andrew Saunders, who was a professor of mine at the SUNY College of Environmental Science and Forestry in Syracuse, New York. In a gap between classes one morning, I walked into Andy's office to find him bubbling with excitement. He had just had the most remarkable experience. He had seen the unseeable! Before coming to campus Andy had put some birdseed in the feeder outside his home, not far from the university. As Andy walked back inside, he passed by the picture window overlooking the feeder. Out of the corner of his eye, Andy saw the slightest rustle in the leaves on the ground, or at least he thought he did. Pausing for an extended moment, he watched.

Allowing his peripheral vision to take a broader inventory of the area made the difference. Though not often in focus, our peripheral vision is well crafted to function in lower light conditions, to spot changes in patterns, and, most especially, to pinpoint movement. After a moment, Andy saw one of the leaves wiggle, then a flash of motion. It was not a mouse coming to gather millet or sunflower kernels; it was a never-sit-still, insect-munching shrew. The pointy-faced insectivore darted out, then disappeared back under the leaf litter, not to be seen again.

Admittedly, Andy's sighting was not a grandiose view of painted dogs initiating a hunt in Botswana or the twitch of a tiger's tail before a lethal pounce upon a chital in India, but if you're going to see those bigger and more breathtaking things, this is exactly the sort of practice that will make it possible. Repeated use of our sensory gifts is paramount. When I ask Andy if he remembers sharing that sighting with me twenty-five years earlier, he says, "No, no I don't remember that at all, but that is my everyday." Observing better begins at home, and you may find that some of those little discoveries can equal, and even eclipse, experiences with larger, more charismatic species.

There's a saying in neurobiology that "nerves that fire together wire together," or, in other words, "Use it or lose it." Our brains grow around the tasks we assign to them. Whether building the muscle memory to find an E chord on a guitar or deciphering the angle of an incoming Messerschmitt, repetition is the way we retrain those perceptual muscles. The next time you push back from your desk and take a deep breath, relax into what you can see around you. Noticing an object on a nearby wall might be a great place to begin, but then shift your awareness to what surrounds it. Note that narrow point of sharp vision, but then take a few moments to alternate your awareness in and out of that point of central clarity into the fuzzy periphery, then back again.

Things are not so well defined in the outer reaches of your field of view, but start experimenting with what you can find in the observable periphery whenever and wherever you go. If you are home alone, there's a good chance you won't see much moving. In other settings, the picture could be stirring with kinetic energy from waving grasses in an open field to pedestrians navigating crowded streets. Why is this important? Envision stepping out of your condo on a workday. Before you even get to the car, you catch the sound of a bird chattering up a storm in an obvious state of distress. Having full command of your visual field might be the deciding factor in finding who is disturbing the peace and subsequently figuring out a little bit more of what makes that bird's world go round.

Practice expanding your perceptive field and you will soon be well rewarded. Maybe, in the static world of the office, a coworker starts spinning a pen in midair to your right, or a fly buzzes in from the lower left. Movement outdoors could be something as small as a single spruce needle falling from a tree and reflecting a glint of sunlight before it strikes the ground.

Art and other focused pursuits help us refine our senses as they hold us accountable to what lies before us. Our attention and recall perform better when we are curious, when we find something beautiful, and most especially when we are striving to reproduce it in a sketch, clay, or some other way. There are endless strategies to discover all that is in the detectable universe. Be creative in your own way as you press your senses back into service.

− 4 −

THE SIGNAL
IN THE NOISE

ANYONE WHO HAS EVER WALKED into a cafeteria full of children at lunchtime knows that the roar of an elementary school eatery can hit your ears like a blast of cannon fire. Yet, if we step into the fray, we often find that the noise the children are making can be understandable, funny, hysterical even, and of course informative. But it requires taking a seat at the table.

As you begin to explore your environs with a renewed sense of awareness, patterns will begin to emerge. The sounds made and the gestures used by animals are anything but random. The dawn chorus, for example, occurs in the morning for specific reasons, including the fact that prevailing winds tend to be lower at sunup; and when still air combines with low ambient temperatures, birdcalls come across more crisply and clearly. Amphibian choruses, by contrast, strike a tune at night. A frog's permeable skin is less prone to dehydration in the cooler air of evening, and being active after sunset helps protect frogs from a plethora of daytime predators. Biology dictates how communication takes shape. Animal behavior, as a result, becomes a window into their world, and the more views we have of this world, the better equipped we are to interpret what animals are saying.

As the students attending my classes begin to take more notice of the purposeful behavior of animals around them, they often ask, "Which is smarter, a buffalo or a bear? A magpie or a mouse?" The answer is always the same—every creature is a genius. They are the world's leading experts in what they need to do to survive. Inevitably, humans, like the prevailing winds or drying heat of the sun, are a part of the world they must navigate, and they adapt to our behaviors just as they would to the behaviors of other animals in the spaces where they live. Their awareness of their surroundings—and of us—can be mind-boggling.

As an example, my wife, Jenny, and I took a work trip into the wintry interior of Yellowstone Park some years ago. Common ravens are often seen near the hotels, parking areas, and points of interest. Animals may seem randomly scattered around your home or the places you travel, but this is not so. In our area, the crafty birds act as though they have invisible watches telling them precisely what time to assemble at a given location to greet wide-eyed tourists each morning. Unbeknownst to these travelers, however, the ravens are not there to see their smiling faces, but to steal their lunches.

Arriving at Old Faithful that particular day, Jenny and I scrupulously secured our belongings with full knowledge that area ravens have a knack for thievery. Even in this wilderness setting, *Corvus corax* has learned how to operate zippers, undo snaps, unzip plastic bags, untie straps, remove tape, and circumvent buckles—and in one case, screw the lid off a pint-sized peanut butter jar—all without the aid of hands. Our human-altered world overlaps with their wild landscape, and some species have learned to take full advantage of what our commingling has to offer.

Following our hour-long meeting, Jenny and I returned to the parking lot and spotted one of the tourists' snowmobiles being ransacked by a conspiracy of ravens. The birds were tearing apart a backpack and had already strewn clothing, receipts, cash, credit cards, and IDs out upon the snow.

"What idiots," I said under my breath. "There's plenty of warnings out there telling visitors how to keep their things safe. Get with the program!"

As we approached, I began to recognize some of the flotsam. These were *my* receipts, *my* clothes, *my* cash, *my* credit cards, and even my driver's license! The resourceful birds were teaching me an important lesson. These bird brains, sometimes disparaged as ebony "dumpster chickens," have faculties that surpass my own. Even when we know local birds have certain tricks up their feathered sleeves, we don't often grasp the breadth of raven resourcefulness until we have such an encounter.

Lynelle Schuerr, a playground aid at a suburban elementary school, had her own encounter with an intelligent bird adapted to what humans were doing in her neighborhood. Every Friday afternoon, the local crows would gather on the elementary school roof and on trees and utility poles near the playground. The timing of the conclave was unclear, that is, until the children were released outside to play—Friday was popcorn day!

As you can imagine, plenty of the snacks fell to the ground and were snapped up by the waiting crows. Lynelle, who lives across the street from the school, noticed something quite interesting one particular Friday afternoon. This Friday happened to be a holiday. Classes were not in session and yet the usual ten to fifteen crows assembled on the school grounds. The crows had timed their arrival not according to whether children were present or cars were in the parking lot, but according to the calendar. It was Friday. It makes you wonder: Had they somehow been counting the days? Want to know when popcorn day is or when the tourists are about to show up? Consider watching the local corvids. Although much of what animals say and do can seem mysterious at first, there are definitely patterns that come out of the apparent chaos.

Anyone living in an agricultural zone might notice an influx of corvids or raptors when it is time for the hay to be cut. The birds are not so much fans of heavy machinery; rather, like the crows that Lynell saw gathering popcorn from the kids, these opportunistic creatures are interested in activity that results in food. Mowing machinery injures, kills, or displaces lots of small mammals, reptiles, and insects that the birds then dine on. Presence and absence can communicate as loudly as words.

Go for a walk sometime soon and look at your surroundings through a new set of glasses. You don't have to take a walk expressly for this purpose either; your next trip to the grocery store or stroll with the dog will fit the bill. Take a survey of what's going on. Make a physical or mental note of who is involved, along with what, when, where, and how things occur. Look for synchronicities. If you do this regularly over the course of a year or more, patterns will begin to jump out at you. Do you notice the owls hooting more commonly at certain times of the year? How does this relate to other things going on, and could there be a connection? Experiment with this contextual matching game as a way to add meaning to the signals you pick out. Humans are pattern-seeking machines, and this trait will help you understand animals when you might otherwise overlook what they are doing.

Have you ever noticed, as a friend did, that all the ducks commonly disappear in the middle of summer? The group of mallards that inhabit the wetlands and lakeshores around her home vanish for a few weeks. And when the birds finally return, the males are nowhere to be found. She noticed that the ducks disappeared when recreational paddlers were also seen chasing geese along the reservoir shoreline. "Strange," she thought. "Why don't the geese just fly away?"

There is a unique, waterfowl-specific signal at play here. Waterfowl, including ducks, geese, and swans, are unusual in that they molt all their flight feathers at once. Unlike other birds, such as the hawk soaring overhead that appears to be missing a single "finger" on each outstretched wing as it undergoes its sequential molt, waterfowl are rendered flightless for the following three to five weeks by losing all of their flight feathers at the same time.

The molt in North American waterfowl takes place between mid-June and early September and varies depending on the species and sex of the duck. The peak of the flightless period for mallard drakes is the first week of July, and for females, the peak occurs in the first week of August. Ducks also undergo a significant energy stress as their bodies regrow feathers, and as a result, they disappear into the hidden corners of wetlands. Safe among cattails and bulrushes, ducks find high-protein green plants and invertebrates to help fuel the renewal of their plumage. In the case of the geese, this friend would get so frustrated with canoeists and kayakers during the molt because they would make the birds run for it across the surface of the water. The boaters didn't understand the compromised condition of the fowl and unwittingly targeted them for a bit of cat-and-mouse fun during outings.

When the ducks finally returned, our friend was also a bit perplexed because all the vibrant green-headed drake mallards were missing—only the darker-brown females remained. Did the males migrate away? As it turns out, each season of the year has unique stories to tell, and for the mallards, the boys undergo a full-body wardrobe change and are hiding in plain sight. Drakes shed not only their wing feathers, but also their gray, black, and green body and tail feathers. Males in their "eclipse plumage" look just like the females. But look closely and you will notice that the beaks of some of the hens are missing their telltale bright-orange color and black markings. The mallards with drab olive beaks, as it turns out, are incognito males.

Put a mark on the calendar for the same day and time next month or next year, and see if what you observe is the same or different. If you have a habit of posting nature sightings online, scroll back through your timeline to see what clues it might contain to seasonal happenings in your area. Want to know when more of these interesting coincidences occur? Have a seat or take a walk and make a note. No matter what your preferred method of observation or whether you end up chronicling what you find, you are sure to find exquisite patterns unfolding from the seeming randomness of nature.

− 5 −

MAKING SENSE
OF MINUTIAE

WATCHING AREA WILDLIFE in the weeks and months ahead, you too will likely uncover behaviors that many people miss. You'll be doing far more than logging factoids. Dynamic stories about your local area will take shape.

Along my own journey to understanding animals' behavioral patterns better, I explored the Brown Farm in Blacksburg, Virginia. The Brown Farm is an old farmstead on the periphery of town that fell into disuse and was later bequeathed to the municipality. I was pursuing a master's degree in wildlife science at the time, and I attended many contentious public meetings over what should, or should not, be done with this idle parcel of land. One faction held that the rolling upland hills should be leveled and converted into much-needed athletic fields for the community's youth. Another group insisted that the farm would be best left as a nature preserve to benefit the sprawling university town.

I joined a handful of other Virginia Tech students and knowledgeable locals to survey the wild plants and animals on the farm. We spent countless hours logging our findings before and after classes, and on weekends. One early March morning, I was investigating the central ridge running through the property. Above me, against the backlight of brilliant sunshine, I could see a swarm of insects hovering at the base of a large tree.

"This is odd," I thought. "It seems a bit chilly for bugs to be out."

The bugs were not spaced evenly across the trees in that stand. They had gathered on one particular tree. Their attention was not on all sides of the tree either. They were laser-focused on the sunny side.

Morning temperatures were close to freezing at the farm, and yet here they were. An entire cloud of cold-blooded organisms had abandoned the places where they sheltered from the chill of the night and were having a field day. It made little sense. Letting curiosity get the better of me, I walked up to that tree and discovered it was a sugar maple. The insects were not any random bug either. They were dipterans, members of the group of true flies not unlike the ones that buzz around your house and land on the kitchen counter to pad about with their dauber-like mouthparts.

Kneeling down on the dry leaves, I discovered an evenly spaced, horizontal line of holes punched through the maple's bark and into its living cambium. I knew this to be the work of a type of woodpecker called a yellow-bellied sapsucker. In the cool nights and warm days of spring, the sappy precursor to maple syrup was oozing out onto the tree's exterior. That's what the flies were there for. The maple was the only breakfast buffet in this patch of woods, and understandably, it drew a crowd. The sapsucker was tapping the sugar maple for its own nutritional needs and was also happily mopping up any invertebrates that became mired in the sugary liquid while they tended to theirs.

What had started as a chance observation of strange insect behavior told the tale of an intricate chain of events linking the lives of a plant, a bird, and some insects all gathered at this exact spot at this specific time of year. Plucking a single, seemingly insignificant thread can lead to seeing the warp and weft of so much else.

Watching insects and other invertebrates, as it turns out, is a great way to discover all kinds of new narratives. Spend a few moments with these abundant and remarkably diverse organisms. Kneel down and lean in if you need to. And if that doesn't quite do it, pick up a hand lens or a pair of close-focusing binoculars. A neat tip if you're in a pinch for a magnifying glass, but have a pair of standard binoculars, is to flip them around and look through one of the lenses backward. Voilà, magnification! Try tracing bark beetle galleries with your fingers to see where the little larvae started their lives, grew fatter, and eventually exited as full-grown beetles. Find an ant trail near a parking lot or in the grass around it—well-used trails can look like miniaturized human hiking paths of packed, bare earth, and you can often find them just beneath the surface of mown grass. Follow it. How far does the ant trail go and where does it lead?

Starting with these smaller subjects is a great way to begin recalibrating your senses and picking out new patterns. Macroscopic discoveries, though small, are deceptively simple and routinely point to larger trends close by. The more dots you connect, the better situated you will be when deciphering meaning from other corners of nature's menagerie. Take a break from routine tasks such as weeding, planting, or harvesting to turn an ear and an eye toward what's winging or scuttling about in your garden. It might help to channel your inner child and remember the days when you could stare for hours at an ant making its way up a blade of grass or follow water striders skittering across the surface of a water barrel. You are never too old to be fascinated by the little things.

Chuck Ernst, a retired optics engineer from outside Colorado Springs, Colorado, began exploring the areas around his home with smaller subjects in mind. In the wake of the devastating Black Forest fire in 2013, many residents looked upon the charred landscape as ruined. Chuck chose to see it differently. To him, things simply looked changed. Chuck became captivated by spiders, including their variety, their dexterity when feeding or spinning webs, as well as the prismatic effects of light on their silken threads. Much of this had previously been hidden or overshadowed by other aspects of life in the unburned forest.

Spiders can produce silk that can reflect, or absorb, various wavelengths of light. Arachnids can consequently make their webs, or portions of them, visible or invisible based on the needs at hand. For example, when a strong but invisible structural support is needed, spiders can produce exactly that. Alternatively, when a visible "landing strip" is needed to lure prey into the center of an orb weaver's net, the web owner can accommodate that need as well. Spiders can even adjust the physical and chemical properties of their silk to vary the degree of stickiness of a web to catch prey.

In one instance, Chuck found himself completely amazed as a spider he was watching—poised on the arched top of a leaf—extruded a single strand of silk from the rear of its abdomen. The strand rose skyward and before he knew it, the spider lifted off into the air. He literally saw it make its own monofilament parachute and fly away. We seldom think of small, sedentary, unwinged animals making large-scale movements, but some, like spiders, have evolved unique superpowers to do just that. A single spider was the only living organism found by scientists nine months after the eruption of the volcanic island of Krakatoa in 1883; it likely traveled from neighboring islands—a distance of perhaps twenty-five or thirty miles. That spider probably would have traveled by air using precisely the method observed by Chuck.

Many of the tiny creatures we tend to look down on and disparagingly call bugs are quite persnickety when it comes to their homes and relationships with other walks of life, especially when it comes to table fare. Being harvesters of much of the world's plant energy, insects go to great lengths to access those resources. This can be no small matter when much of that energy is locked up within cellulose and lignin cubicles, and protected by complex chemical firewalls. As a consequence of jumping through a plant's defensive hoops for so long, many insects have become inextricably linked to specific plants, and to those plants alone. This gives you an opportunity for some up-close ecological research in your own backyard and explains why the leaves on the gooseberry bushes you have planted in your vegetable garden can be decimated by sawflies while the leaves on the neighboring blueberries you planted are untouched: the larvae of the spotted gooseberry sawfly are single-mindedly focused on the former and completely ignore the latter.

Insect abundance affects life further up the food chain and impacts the prevalence of mammals, birds, reptiles, and amphibians. If there is no food for moths and butterflies, there will be far fewer insect-eating birds like warblers, tits, and chickadees. Even seed-eating birds will be less abundant, as they, too, raise their young almost entirely on invertebrates, most notably caterpillars. Not only will this eye for the diminutive expose a wider range of informants, it will also make identifying changes in what is normal, and hence, revealing messages, much easier.

Along these lines, consider giving plants and pollinators some breathing room during crucial times of the year. In suburban and urban environments, pollen and nectar sources can be in short supply in spring. A yard maintenance hiatus or choosing to leave one segment of the yard unmown can be helpful to both plants and insects. Another side benefit is that a break also greatly reduces the amount of time, expense, noise pollution, and petroleum used by homeowners. Watching the grass grow can be a win-win opportunity on many levels.

My own observations of insects led in some interesting directions in a less-than-urban setting. Some years ago, I gave myself a full summer of just noticing butterflies in the wilds of Yellowstone country. I say "noticing," as I was not studying them per se; I had simply made a conscious decision to open the door to them. Clicking my sensors over into butterfly mode completely changed the way I experienced the surrounding areas. During regular outdoor tasks, I would make passing mental notes of the different colors, sizes, shapes, and numbers of various butterflies. Along the way, I developed not only a sense of the locations where they might be found, but also a sense of who they hung out with and what flowers they visited.

I also acquired a newfound appreciation for the manner in which they moved. Without really trying, I found I could identify certain species from a few feet to hundreds of yards away without the need of binoculars. Some lepidopterans have bold markings that make recognition easy, like the yellow- or creamy white–edged wings of deep-maroon mourning cloaks or the bold yellow background color and black stripes of swallowtails— there is nothing else like them. But I also began to realize that many species could be picked out from across a field simply by the way they fluttered. You may have had a similar experience recognizing a friend from a great distance purely by the way they walk or how they gesticulate with their hands in conversation.

The flight pattern of the Hayden's ringlet, for example, has a slow, floppy quality, which differs from the flight pattern of the similarly sized and colored gray-brown small wood nymph. The latter, by contrast, flies in sharply erratic up-and-down movements. I sometimes snicker when I'm hiking and see a small wood nymph. They strike me as over-the-top, fatalistic drama queens of the butterfly world. They seem to be saying, "Oh, I've fallen. Oh no, now I'm back up, but I'm collapsing over here now. See me flashing to this flower? Oh no, but I've fallen again. Oh, but I'm fine... Now I'm off!"

Those erratic flight paths, as it turns out, are an exceptionally good strategy if you want to avoid being eaten. Rabbits make use of this zig-zag movement when escaping danger, and squirrels often wave their bushy tails one way while moving their bodies in another. Like these mammals, the wood nymph's erratic flight makes it much harder to catch.

By the end of that summer, I had developed such a sensitivity to tiny flying things that there was no way I could *not* see them. My sensitivities to different flight patterns extended beyond those of wood nymphs, and into aspects of butterflies that were difficult to articulate with words. I began to notice that light levels and stages of flower blooms affected their behavior. My hypersensitivity to winged insects and the way they moved sometimes distracted me from conversations with other people. If the specimen was distinct or unusual enough, I occasionally had to politely step away to have a closer look.

Some might say, "I'm not into bugs," but you may find, as I did, that these jewels in flight can point out other items that do interest you. In the 1990s, I coordinated field activities for a dietary study of coyotes in the Adirondack Mountains of New York state. Each month for an entire year, I mobilized volunteers to cover thirty miles of logging roads in search of coyote scats. Paradoxically, one of the greatest aids to finding the best and freshest samples came from butterflies. If you have ever seen a group of butterflies gathered on the edge of a mud puddle, on a carcass, or on a patch of urine-soaked ground, you can be sure of one thing—they are all males. Male butterflies need to replenish essential salts, primarily sodium, throughout the breeding season. Conveying packets of sperm to females for weeks on end translates into a steady loss of these salts and thus requires the males to replenish their saline supplies through this puddling behavior. As repulsive as it might be to watch a butterfly unfurl its elegant, coiled proboscis to soak up the juice from a turd, you are witnessing it perform a vital life function. Since male butterflies, like male birds, are generally more colorful than females, it makes spotting scat samples easier, thanks to the colorful, fluttering halo surrounding them.

You can explore these invertebrate behaviors in a local park or in your own garden. The dots and connections will continue on into infinity. And when the national media addresses issues such as the loss of pollinators or global insect decline, you will have a starting point to consider the topic more directly. Are the bugs that you have been watching directly affected by some of these global patterns? Do you see the same kinds of insects over and over, and are they doing better in your town compared with elsewhere? These and so many other questions will emerge intuitively as you observe more closely.

The Brown Farm in Virginia, where my bug, tree, and bird experience occurred, was ultimately designated a nature park. Named the Brown Farm Heritage Community Park and Natural Area, it is now maintained as a wild, open space. Although I have not been back since its status was updated, I'm told that a small parking area, a pavilion, and a footpath have been built. The landscape remains largely as I remember it—full of life for others to discover.

— 6 —

ANIMAL SENSES

ECONDHAND KNOWLEDGE GAINED through animal senses can be one of the greatest gifts to learning what is going on around us. Making use of the constellation of animal superpowers elevates our awareness severalfold. Even if we can't see into the ultraviolet spectrum like an insect or respond to weak electrical impulses like a shark or feel fishes' hydrodynamic trails like a seal, the behavior of other species gives us clues to what is going on in the more-than-human world. Although human olfaction is geared more for sniffing mangos and bananas than for huffing fumes off a drowned bison, as a bear does, for instance, we can ascertain the presence of a hidden carcass by closely watching a bear as it sniffs its way around an icy pond in early spring.

When teaching field-based classes, I sometimes motion with my arm as if waving an imaginary wand. I tell students they have been transformed into bison, coyotes, or harlequin ducks. Participants are then tasked with figuring out how to abide by unseen territorial boundaries, where to select the best grazing sites in a valley under several feet of snow, or when to set off on their continental-scale migrations. How would you navigate without a

GPS, map, calendar, or compass? Where should you go and when would you leave? And how will you know when you get there?

Brief as it is, this thought experiment reveals the rift that exists between humans and the rest of the animal kingdom. Making a living in the wild is not child's play, and without a trove of learned and inherited knowledge, animals wouldn't know where to show up for breakfast, where to sleep, where to get a drink, who to trust, and who to avoid, and would probably end up being someone else's lunch.

To survive in the wild requires superpowers. Some pet owners train their animal partners to hone the abilities they inherited from their wild forebears. Animal companions can understandably give us some idea of the depth of awareness available to their wild cousins and act as useful bridges between what wild and domestic messengers are telling us.

Bound for the mailbox one Saturday morning, my son, young George, and I have a chance encounter with our friend Colette Daigle-Berg. Colette, a former Park Service law enforcement ranger, is conducting search-and-rescue training with her female border collie–yellow Labrador retriever mix, Chapter, and Colette asks us if we would help with the day's exercises. Although our formal introduction to Chapter is still a few hours away, we happily agree. Chapter's name reflects the pup's role in Colette's life, as she embarked on her next life chapter following her retirement from her Park Service career. For our part, George and I are to create an interference pattern, an olfactory red herring, over a scent trail laid down by Colette's assistant two hours earlier. In short, we are to complicate an otherwise simple path of odors and make things a bit more challenging for Chapter.

Though we are largely oblivious to it, each of us exists in an ever-present cloud of minute fragments that emanate from us like a microscopic ticker-tape parade. Our unique clouds are partly composed of the roughly two hundred million skin cells we exfoliate each hour. We all resemble Pigpen, the character from Charles Schulz's *Peanuts* cartoon with the persistent plume of grime billowing around him—as it billows around all of us, all the time. The chemicals in our sweat, combined with shed hairs, textile fibers, molecules of soap, detergent, and perfume, create a personal scent signature. This odor is a physical extension of our being, and we are leaving a lot of ourselves lying around for someone, or something, else to follow.

Chapter recovers avalanche casualties, drowning victims, and lost hikers for her job, and she depends on the invisible flotsam we generate. At times, the scent particles operate exactly like, well, particles. We might imagine them as minute sticky notes pointing the way. They adhere to the ground or to upright objects such as bushes, buildings, and fences. In other cases, odors behave more like fluids that eddy in, around, behind, and even through some of those very same objects. Scent can pool in low-lying areas such as a pit, a ditch, or a river bottom, and is further altered by each passing breeze.

A couple of hours after young George and I make our contributions to the smell-o-scape by hiking along in normal fashion, Chapter navigates the entire one-mile challenge course with ease. This was a walk in the park for her. Colette later shares with me a single black line on her GPS unit that shows where her human assistant walked through rolling hills and sagebrush, and back down to the road. Colette then shows me a blue line: the path that Chapter took. Chapter's line lies atop the target person's path almost exactly. There is a slight bobble at the beginning, where Chapter had to tease out our odors from that of the person she was supposed to find, but then she was off to the races. The only other discrepancies are periodic diversions of a few feet downwind. Chapter also cut off a wide, misleading tangent of about two hundred yards from the end of the exercise when she saw Colette's assistant standing on the roadside; this, too, is part of the training. In the case of a person teetering on the edge of life, search-and-rescue dogs are taught to leave the scent trail and go straight to the individual whenever possible.

As Chapter bounds down the dirt embankment and onto the road, young George erupts with a resounding, "Yay, Chapter. You did it!"

None of this would surprise anyone who has worked with scenting dogs. This is routine stuff. What Colette shares with me next, however, really blows my mind.

Three years before our training run with Chapter, a man parked his car at a trailhead in Mesa Verde National Park, Colorado, and never came back. Colette and several other dog handlers were asked to canvas a specific area in hopes of finding evidence of the missing man. They, too, failed. The hiker's remains would later be found, based on an anonymous tip, in a spot far outside the established search area. During the search that did take place, Colette said Chapter found something far more unexpected than a lost hiker.

At one point, Chapter began signaling that she had found human scent near the roots of an overturned tree. The dutiful pup did a perfect "refind," as Colette calls it—a trained response where Chapter places her front paws up on Colette then returns to the source of the scent and lies down next to it. This signal is the dog's way of saying, "Look here, right *here!*" Colette could see nothing.

Moving closer to the exposed roots of the tree, Colette still found no evidence of human remains. It seemed as though Chapter might be mistaken for once. And when it appeared that her mom might lose interest and drift away, Chapter did something she had never done before in a situation like this—she barked. Chapter's outburst was so odd that Colette took out her phone and shot some video footage. She shows me the clip as we talk about the encounter. Still feeling a bit unsure, Colette praised Chapter, rewarded her with a treat, and moved on.

On a hunch, Colette checked in with one of the resource specialists at park headquarters afterward. "Are there, by chance, any Native American burials in that area?" she asked. In a guarded way, the specialist answered in the affirmative. "Yes, yes, there are. We don't typically talk about them, though."

What made Chapter's find so remarkable was that this "lost" person was not from a few years before, or even from this century. This area holds ancestral Puebloan burial sites that date back between eight hundred and twelve hundred years. Chapter was not sensing the recently departed. She was mingling with the ancients. What would it be like to go about your day with the odor of the ancestors from centuries, if not millennia, ago swirling around in your nostrils? The thought is nothing short of dizzying.

Dogs trained to smell cadavers focus on a cocktail of fatty acids emanating from decomposing flesh and bone. Dogs and other species that are hyper-focused on these chemicals can often detect them in concentrations of as little as one part per trillion. To get a handle on what that means, envision dissolving a tablespoon of sugar in a volume of water equivalent to two Olympic-size swimming pools and still being able to taste the sweetness—that is how good they are.

Chapter has put these same skills to use when locating drowning victims beneath dozens of feet of water, and in one case pinpointed a deceased man beneath a foot of river ice: human crews couldn't locate the victim until they used ground-penetrating (in this case, ice-penetrating) radar. I have always maintained that if we could smell as well as our canine friends can, we would probably be down on our hands and knees sniffing the lawn, rocks, and fire hydrants right alongside them.

Lauri Travis, an associate professor of anthropology at Carroll College in Helena, Montana, has taken advantage of the powers of those canine noses in her paleontology and archeology work. Dogs are commonly used to find drugs and buried bombs, and even to detect estrus in cattle. Lauri is among a small group of researchers harnessing canine noses to learn about the past. Lauri's work started in 2018 with Hannah Decker, an undergraduate student in zoology at Carroll, and a shelter-rescued border collie–Australian shepherd mix named Dax. At that time, Lauri was working on a ten-acre study site and was permitted to dig no more than three one-meter-by-one-meter test pits to search for artifacts. Their teamwork paid off, as Dax's nose was able to help narrow down a vast field of possibilities; it would have been a multi-acre roll of the dice otherwise. Dax's oldest find to date has been a five-thousand-year-old bone buried beneath fourteen inches of soil. No man-made device has yet been able to duplicate such feats.

Imagine extending this sort of lesson from domestic super-sniffers to more of our local wild informants. What else might we be able to find? Entire books are filled with examples of these super-senses from across the zoologic spectrum, and as I write these words, more are being discovered. Elephants decode long-distance, low-frequency vibrations through their feet. Certain migratory birds are probably able to see Earth's weak magnetic field and use it as a navigational aid. Mice are crooning to their darlings in ultrasonic frequencies. Dolphins are reflecting sound off nearly identical objects to differentiate textures and materials. Eels are able to both stun prey and titillate their mates with their electrical pulses, and the list goes on.

Planet Earth is literally brimming with a faunal cornucopia of sensory talents. Who needs Marvel Comics when the non-humans living in your own house or flowerbed can manage parallel feats? Despite our nifty gadgets and techno smarts, we are no match for biological evolution, which has been picking the winners and killing off the rest across millions of years of research and development. When your life is on the line and dinner is not waiting on the table at the end of a long day, every detail counts. For a lizard that misses the shadow of a hawk slipping across the ground, a butterfly that overlooks the fact that the sun has disappeared behind mounting rain clouds, or a young male deer oblivious to the approach of a larger adversary, the consequences can be severe. Meeting your maker sooner than planned is a real possibility.

We humans have put a lot of effort into our game of one-upmanship with the rest of Earth's biota, but more than anything, this one-sided view has stifled our ability to see what is actually there. Watch your cat or your parakeet, or the resident opossum or groundhog. Wait until you see it react to some kind of external stimulus. See if you can tell what brought its head to attention. What made it rotate its ears? What elicited a call? Look in the direction it is looking, then listen carefully for what it has heard. Draw air slowly into your nose on the chance that you can catch a whiff of something it found interesting.

I did this with wild turkeys each spring for a period of three or four years, logging each sound that triggered them to gobble. The final count of gobble-inducing noises climbed to over a hundred natural and man-made sounds. The noises ranged from overhead jets and barking dogs to banging car doors and thunder. The diversity of that list speaks to the level of attention turkeys have for their surroundings at all times. Not all of what they noticed was a matter of life or death, but to truly get to know their home, they needed to take it all in. And by watching them and making a note of the stimuli they were responding to, I not only got to see things I would have completely overlooked, but also began, finally, to see the world a little more closely from their point of view.

In a somewhat comical version of secondhand knowing through animal senses, we were sitting in the living room one evening when our Labrador Hobbes suddenly leaped to his feet and uttered a menacing growl in the direction of the kitchen. I rose to my feet to find he was focused on something in the vicinity of the stove. Was Hobbes cornering a phantom menace that was about to pounce? I soon realized that our trusty defender had finally spotted the fresh loaf of bread that Jenny had pulled out of the oven and placed on top of the stove to cool. Good boy, Hobbes.

— 7 —

OUR SENSES

BEFORE YOU RESIGN YOURSELF to the idea our senses are inferior to those of other species, I'm happy to report that all of us have some pretty outstanding abilities of our own. When we are looking at appropriate things, sniffing the right stuff, there are times when we can put other species' super-senses to shame. The odor of fruit or rancid oil, for instance, are both scents that we humans are adept at recognizing, even in concentrations of parts per billion.

In my experience, however, instead of celebrating their abilities, people often focus on perceived shortcomings they worry will limit their opportunities to learn from nature. A parent might confide in hushed tones in the lead-up to a class that their child "cannot sit still very long" because he or she has ADHD. Older individuals might report that they have a tin ear, or that their eyes aren't as good as they used to be, or that their nose is bad. However, sensory challenges are not the barrier to experiencing the wider world that we think they are. In nature, there is a place for everyone, even if you have macular degeneration, ADHD, Raynaud's phenomenon, or a hip replacement. So-called obstacles may be unseen doors to enhanced experiences.

During my early twenties I realized something was amiss with my hearing. I remember walking across the room and tapping on the CD player. There was an apparent malfunction. I was learning the diverse songs of eastern wood warblers. The narrator introduced the next bird recordings by saying, "Blackpoll warbler," but nothing played. Only after I knelt down with my ear next to the speaker and turned the volume up did I hear what I was missing—the high-pitched trill that is the blackpoll's song. A trip to the audiologist confirmed what I begrudgingly suspected: I had hearing loss. The problem was mostly in my left ear—another downside to being a right-handed marksman, compounded by years of power tool use during various home projects.

While writing this, I played a hearing-test video on YouTube and found my current range of hearing starts around 70 hertz (Hz) on the low end. In the higher frequencies, my left ear started losing ground around 8,500 Hz—well below the average bandwidth of 8,900 Hz and higher used in blackpoll songs—and though my right ear was a bit better, sounds above 14,000 Hz were inaudible. A young child with a full range of hearing may hear up to, or slightly over, 20,000 Hz.

Despite being barely into my third decade of life at the time of the blackpoll incident, I was told that my hearing was equivalent to that of an elderly man. No one had ever told me about the risks of all that noise. In addition to dealing with the hearing loss and the ringing tone of tinnitus that stepped in, I had also worn glasses since the age of twelve. The extent of my compromised senses left me crestfallen, but the predicament was not as dire as I thought.

I focused more intently on the frequency ranges I could hear despite my limitations. Before long, I was identifying birds by ear that other listeners overlooked simply because I was more attentive in my listening. If you are like me and can't hear certain birds or cannot see that swarm of midges hovering over a river without your glasses, you might find yourself pleased to discover that different things present themselves to you. As an example, I would later find, as did a friend with similar hearing loss, that I noticed the low rumble of distant cars on dirt roads long before others with normal hearing did. As a result, we had more time to get the dogs under control and out of harm's way.

Individuals who are hard of hearing may benefit from enhanced peripheral vision. People with limited sight often have improved memory for sounds. Those with red-green color blindness can detect camouflaged objects better than their normal-sighted peers. The camouflage-busting ability of color-blind servicemen was used by British and American forces to spot hidden enemy encampments during World War II and the Vietnam War. If you happen to be looking for a cryptically patterned grouse nesting on a forest floor, your color-blind friends might be better equipped to spot her than you.

Thinking of ourselves as being in some way dis-abled directs our attention to the things we cannot do, rather than to the things we can do. In many cases, individuals viewed as being in some way handicapped exceed the rest of us by a long shot in other areas. Shane Doyle, a Crow tribal Elder, framed the matter nicely during a program for young adults with autism. "There is no such thing as handicapped in our culture. Everyone is *alternately able*." And indeed, Crow society not only embraces but celebrates diversity in all its weird, funny, and awkward manifestations. Getting the most out of your time eavesdropping on animals has less to do with what you can hear, see, or smell, and more to do with what you *do* with what you can hear, see, and smell.

The late E. O. Wilson had an influential career in ecology at Harvard University, studying ants—which won him the Pulitzer Prize. But he might never have embarked on that career path had the spine of a pinfish not punctured his right eye during a fishing accident when he was seven. While the injury closed one door for Wilson—he would no longer be able to study colorful and highly mobile birds and mammals—it opened up another. A career studying the smaller and less appreciated subjects was just waiting to be made.

The American paleontologist Jack Horner attributes much of his MacArthur Fellowship, aka Genius Grant, success to his dyslexia. Labeled "lazy" by his father, Horner failed out of virtually every class he ever took and never earned a single degree. He forged on despite what others considered insurmountable obstacles, and went on to find the first dinosaur embryos in the Western Hemisphere and evidence for parental care in the extinct reptiles. Horner has since earned widespread recognition and honorary degrees, and most importantly, he has succeeded in finding what everyone else had been missing.

Temple Grandin, the American scientist and animal behaviorist, attributes much of her ability to understand nonhumans to her autism. Grandin says that she thinks "in pictures, not verbal language." She notices things that neurotypical people walk right past. As she states in her TED Talk, "An animal is a sensory-based thinker, not verbal. [It] thinks in pictures, thinks in sound, thinks in smells." In her opinion, "language covers up the visual thinking we share with animals." Given how many geniuses of the past are thought to have possessed autism-spectrum disorders—

think Sir Isaac Newton, Emily Dickinson, Albert Einstein, Michelangelo, Mozart—Grandin believes too many kids with Asperger's and similar traits are being sidelined and "not ending up in Silicon Valley, where they belong."

A child who can't sit still during math lessons may be perfectly suited to discovering mathematical connections plucked from nature. Bullet trains have been made faster and quieter by using designs modeled on the beaks of kingfishers, and jets have become more stable by borrowing ideas from the supple wings of hawks. A restless child might grow up to create great works of art inspired by haystacks or pond lilies, like Claude Monet, or be moved in the way composer Ludwig van Beethoven was when he incorporated the sounds of birdsong into his Sixth Symphony.

In earlier times, diversity was not scrutinized, but utilized. Many who did not fit social norms became our best artists, hunters, spiritual leaders, problem solvers, and intrepid explorers. Firing on all our sensory cylinders has a way of elevating the whole of society, but only when we celebrate the full range of talents. We all have our superpowers.

Whatever your talents for observation are, they will all be for naught in the absence of one thing—exposure. The more time we lend to our surroundings, the more novel realizations appear. The mind makes use of many pathways to knowing. Consciously, our minds log forty to fifty bits of info per second. By contrast, the unconscious is processing upward of eleven million bits of data in that same time interval. Spending time with someone or something gives our passive data collectors a chance to work their magic.

Some people, for example, can perform quite well with a lesser sense of smell, purely as a result of hanging around unique places and things. One acquaintance became acutely aware of the bewitching scent of porcupines after raising one as a boy. A porky's body odor, as he describes it, is reminiscent of "earwax and popcorn [and] is unmistakable." Forever afterward, he could sense when a prickly one was near by scent alone and know it was time to keep his yellow Labrador close. If he ignored the olfactory clue, he and his father risked spending several hours tending to the dog's mouth full of quills. His sense of smell was not that remarkable by common measures; rather, it had been refined through exposure to detect a distinct odor in a memorable way.

To Jenny and me, the breath of buffalo in Yellowstone in August smells of marijuana smoke. When field biologist and University of Colorado professor Joanna Lambert smells Burger King food in the jungles of Costa Rica, she knows to be vigilant, as this means a herd of potentially dangerous white-lipped peccaries are around. To her, these gregarious, piglike mammals "smell like burgers and fries!" Rattlesnakes have been reported to smell like tomato plants, and the scent of foxes—according to a commenter on an online forum—hints "strongly of sulfur or burnt fried eggs." Danielle Whittaker, author of *The Secret Perfume of Birds*, notes that auklets possess a tangerine scent, and that her beloved juncos have a "forest-like smell of dirt and leaves." The scent of brown-headed cowbirds, her favorite bird, reminds her of "fresh sugar cookies."

Carol, a librarian for a media company in California, honed her senses, and especially her sense of smell, at her local zoo. Carol volunteered time observing the chimpanzees, wolves, cougars, otters, black bears, and condors at the zoo, but it was the elephants that captured her heart. When volunteering for the

elephant monitoring program, Carol collected data on the interactions of individuals and their use of the outdoor space. Time
budgets were recorded for grazing, dust bathing, socializing, and
browsing on shrubby foods in the elephant-friendly enclosure.
Given that elephants have an active nightlife, Carol also volunteered for the graveyard shift.

Nighttime studies done after public visiting hours allowed
for a different kind of experience with the gentle giants. Night-
vision binoculars were provided so she could peer through the
veil of darkness, but Carol felt they scorched her retinas and
made it difficult for her to see afterward. Carol would put down
her binoculars, stash her data-logging tablet, and settle into her
other modes of observation. The nocturnal soundtrack, as it
turned out, was quite revealing.

"They make different kinds of sounds," Carol tells me. "We
have one female who does this almost purring-like sound when
she's content and eating. It sounds like a giant cat, and she's the
only one that does that." Carol says she could hear that sound
alone and think, "Okay, I know where Maggie is." At other
times, with Maggie purring in the foreground, Carol might hear
branches breaking elsewhere in the enclosure and think, "Well,
there's only two of them out tonight. That's got to be Britta."

There were other instances in the inky darkness when Carol
could sense the presence of something large and close, despite
not hearing or seeing it approach. It was as if the elephants in
the room—all three-to-seven tons of them—were standing right
in front of her, and they often were. Despite their incredible size,
the soft pads of an elephant's feet allow them to move in virtual
silence. Maybe she would pick up on the fact that other small
sounds were now somehow blocked or absorbed. At other times,

the airflow might seem to be altered, but inevitably, Carol would recognize a unique scent that cried, "Elephant!"

What funneled into Carol's nostrils was a "musky, hayey, dusty, awesome kind of smell," as she calls it. The mélange of fragrances transported her back to her childhood like an olfactory time machine. She had grown up with horses and sheep, and the elephants made her feel nostalgic for the landscape of her youth. "I love the smell of a barn," she tells me. "The elephants reminded me of that hay-barn smell. There's nothing else that smells like an elephant to me."

Carol eventually tallied more than sixteen hundred hours of elephant viewing over fifteen years at the zoo. With all that repeated exposure, the scent of elephants was deeply ingrained into Carol's sensory pathways. She became calibrated to unconventional ways of detecting elephants, especially when it came to Bart. Standing ten feet, eight inches tall and weighing thirteen thousand pounds, Bart was the only male out of the three elephants at the zoo. Generally a pretty cool guy—"a charming, silly goofball," as Carol likes to describe him—he could get a bit crabby during the musth, a reproductive phase when the testosterone levels of male elephants skyrocket to sixty to a hundred times their baseline concentration. Bulls ooze liquid from the temporal ducts on the sides of their massive heads, and their behavior often becomes unpredictable and dangerous. Bull aggression during the musth has led to injury and even death for many elephant handlers, rural villagers, female elephants, and even black and white rhinoceroses.

When Bart's inner cocktail was brewing, there was one thing about him that was worth noting—he stank, and bad. Although Carol says she could not distinguish the females with her nose, it was another matter when it came to Bart. Carol could definitely separate the girls from this periodically rank-smelling boy. I ask Carol to describe the odor of a musth bull and she says flatly, "It is acrid." When pressed for details, she adds that it smells "kind of greasy, a little sharp, urine-y, and a little skunky." She tells me you can smell it from a great distance and says that many times, when she stepped out of her car in the parking lot of the zoo, she instantly smelled Bart in the enclosure a quarter of a mile away.

As Carol went on to discover, her skills at elephant-detecting extended beyond the confines of her local zoo. An avid eco-traveler, Carol and her husband often vacationed in unique destinations around the globe. Their adventures took them across North and South America, Asia, Africa, and beyond. On safari near Lake Manyara in Tanzania, Carol was being driven along a dusty road as she looked for elephants. "In that dust smell there was this other smell. I could smell that we were getting close." Reflexively, she said to the driver, "Stop, there's elephants!" Carol caught her guide off guard, but he brought the vehicle to a halt mere seconds before a whole group of elephants stepped out of a forest and walked right in front of them. Carol replicated this feat while assisting with a wild dog study in Thailand. Although they couldn't see the elephant, Carol's nose said it was there. She and her companion—who was understandably skeptical—pressed on. A short distance ahead, they found fresh tracks, crushed plants, tail hair, and a powerful puff of that unmistakable scent of a bull in musth.

Carol tells me her best experience was with pygmy elephants in Borneo. She was in a boat on the Kinabatangan River just as dusk was descending, and the guide told them there was a group of elephants in that area. Much of the landscape in Borneo is heavily developed, but small, protected strips of forest border the rivers. These refugia, with their dense stands of trees, are the favored habitat of pygmy elephants, and the animals often wait for people and tour boats to leave before stepping out into the open. The next part of the story is where the energy in Carol's voice elevates a notch or two.

"So, we're coming down the river and it was starting to get a little dark and [the guides] were thinking we were never going to see them, and then... I could smell it. I suddenly got this breeze that went past, and I instantly knew—elephant!"

Despite raised eyebrows from the guides and chiding from her husband, who admits to having a poor sense of smell, they followed Carol's nose. As their boat came around a bend in the river, they could see, not the pygmy proboscideans themselves, but movement in the vegetation.

"You'd see these branches waving and then suddenly it would be ripped down and [then] we heard a little blast of the [elephant's] trumpet!"

What a thrill this must have been for Carol to take those passively absorbed lessons at the zoo out to some of the most exotic destinations on the planet. In the end, they never did see the pygmy elephants of Borneo, but sometimes picking up all the signs can be just as thrilling as seeing the real thing.

You don't need to be an elephant whisperer to extend your olfactory abilities. Although the English language has a limited number of descriptors for odors and flavors, try your hand at describing them anyway. Scribble the results in a notebook or journal. Tackle a description of the scent of a puppy, wet soil, the subway tunnel, or the recycling center. Maybe take a walk through a pet store, pause in different aisles next to different animals and products; draw whatever conclusions and comparisons you can. Take it one step further and ask a friend to lead you through the store by the arm with your eyes closed and see if you can figure out your location. Record these impressions in whatever form makes sense to you, as they have an uncanny way of cropping back up in unexpected places.

You can also try this at home: navigate your house or, better yet, someone else's house by smell. Let your nose guide you. Label the rooms by the odors you find there. The kitchen might come to be recognized as "moldy peach," the closet might be dubbed "old leather," and the laundry room could be the epicenter of "fruity, soapy stuff." Take this approach outside and you may find that you are able to smell the salt marsh much farther away than you previously thought. You might realize that it is trash collection day without ever looking at the calendar. Maybe the big epiphany is that the chocolate factory in that part of the city is still in business!

If you feel like your sniffer is pretty good, turn this exercise around and focus on one of the other receptors. Carol combined scent with sound. How might that work for you? Could you create a map of where you hear a certain sound coming from? This sound-mapping exercise can be an asset in linking sounds with certain aspects of the landscape. Start by scribbling down major landmarks such as streets, houses, and water bodies, and then start filling in where your sound observations are coming from. Add secondary landmarks as needed to clarify certain locations. Even if you don't know the identity of the sound-maker, give it a try. Drafting something on the back of a napkin or in a notebook might help connect additional dots. At what time of day do you hear these sounds? How far are you from a paved road, the library, hedges, or the duck pond? All of this can matter a great deal. Some go so far as to record their notes with mapping software, but the level of high or low tech does not matter; what counts is that you are using your senses.

Maybe taste is a sense you'd like to cultivate more, or perhaps you'd like to work on a greater sensitivity to tactile sensations. Set aside time each week or one location that you frequent, where you intentionally turn on these underused skills. Bear in mind as you work on expanding your sensory capacities that we may not be limited to just the five senses. Michael Cohen, an adjunct faculty member at Portland State University and author of *Reconnecting With Nature*, estimates that there may be as many as fifty-four senses. Our bevy of detectors overlap, synergize, and override one another from time to time. The more senses you cultivate, the more synergy you will find.

Carol's time at the zoo sensitized her to things that otherwise slip past most people's untrained, unaware, and unexposed senses. If we cultivate repeated encounters with new stimuli in familiar haunts—while also setting aside our beliefs about what we can and can't do—there is a lot we might find. It is empowering to know that each of us has our own innate but unexpressed superpowers lying in wait. Even though your senses may not be as sharp as they were when you were a kid or before you became ill, they are still better than you think. Breathe easy and immerse yourself in the possibilities.

– 8 –

LEARNING
BY DOING

HUMAN BEINGS LEARN by imitating sounds and behaviors. As kids, for example, we grow up making noises. Want to learn to speak Italian or Mandarin? You will have to start by babbling and, yes, even butchering those new vocabularies and pronunciations until you gain proficiency as you build new neural pathways. Whether you are on the younger or older end of the spectrum, you might remember a time when you delighted in imitating the sounds of others, especially when it pointed out things you might otherwise have missed.

Following an online presentation I gave, one of the participants shared a story about the goats on her family's farm. When the girl began imitating the sounds their nanny goat made, she discovered that the mother goat had a different call for each of her three kids. After a while, the girl found that she, too, could call any one of the three kids to her side whenever she wished. Unfortunately, the mother goat died unexpectedly. Sometime later, the girl called out to one of the small goats using the mother's "name" for the young one. Convinced that her mother had returned, the jubilant kid gamboled her way. With its tail in the air and head held high, it found that it was not her mother after all, but a little human standing in her place. The woman remembers feeling intensely sorry for the goat when she saw how disappointed it was.

I found my own imitations of animals handy not only out in the woods, but also in the university classroom. When I was tasked with memorizing frog calls in herpetology class, sounds of furry subjects in mammalogy, or birdsongs in ornithology, I discovered that imitating the animals' vocalizations helped me remember them. When I followed along with audio recordings using my own voice, I found those sounds stuck with me better. Even if the name associated with the call was still elusive, the all-important recognition of "Ah, I have heard that before" remained.

Copying animal vocalizations can also be a helpful aid in research and education. Algonquin Provincial Park staff in Ontario, Canada, historically conducted public wolf-howling sessions as an outreach activity. Park rangers howled in hopes of getting a reply as visitors quietly listened. If attendance numbers are an indication, these events were popular. The park's wolf gatherings averaged eighteen hundred attendees and 450 cars per event, and in one instance, the activity drew over twenty-seven hundred people to a single evening's howl. As visitors to the Algonquin wolf howls no doubt experienced for themselves, sound is the vibrating energy of the environment surrounding us and is directly connected to our fight-or-flight defense and pleasure centers.

Fred Newman uses what he hears as a tool to tell stories, and he makes his living by reproducing it. Some may know Fred's work as the artist who portrayed the weasels' voices in *Who*

Framed Roger Rabbit. Fred also scared the tar out of my younger cousins as he imparted the devilish voice to the villain Stripe in the movie *Gremlins,* and he spent nearly two decades as the sound effects man on the public-radio program *A Prairie Home Companion.* I had the pleasure of connecting with Fred through a mutual friend who knew of our shared interest in animals, sound, and mimicry.

Born with dyslexia and not mastering reading until the fourth grade, Fred has always made sense of the world through his ears. "Sequences and order were difficult for me," he says. "But sound is linear and time-based—it ordered the world for me and showed me my place in it."

To aid his work and to find relaxation amid his busy days, Fred has developed a kind of spontaneous meditation. Several times each day, he steps away from what he is doing for what he calls "listening breaks." Fred hits the pause button for a couple of minutes—sometimes as many as fifteen—and just listens. He might take his break while standing outside or when gazing out a window or at an interior wall. The important thing is that he clears his mind and opens his ears. Fred views these breaks as auditory "palate cleansers" that make room for him to experience more of his surroundings.

"Listening is my *now,*" Fred tells me, and he shares the words of a wise mentor of his to reinforce the point. "Fred," the teacher said slowly, "the greatest gift you can give someone is your presence." And our presence, by way of giving our full attention, comes most forcefully into the moment when we take time to listen.

At this point in our conversation, Fred segues into a series of powerful wingbeat sounds formed by forcing air through his lips as he moves his arms up and down. He becomes a bird, a large bird, with the air moving in and out of his lungs trans-forming into the wind beneath those commanding wings. Fred has invoked a new type of presence. He goes on to explain that within the last ten years he has realized an interesting quality in himself and others.

"We don't think with our brains," Fred says. "We think with the totality of our physical being. Move with the sound—we feel it all over."

Woodpeckers are wonderful "starter birds" if you want to get this experience of moving with sound and having it excite all your circuitry. This works because woodpeckers combine sound and motion when they communicate. There is a woodpecker that loves pounding on the plastic hardware of a utility pole near our house each spring. The resonant tone the woodpecker produces bears an uncanny resemblance to the hollow, pulsating ringtone of a British telephone. Each species of woodpecker has a learned percussive pattern that is their surrogate for a songbird's tune. If the woodpecker doesn't get the rhythm right and its call doesn't connect, it won't claim a territory and it won't get a date.

In North America, downy woodpeckers give fifteen to twenty-five beats lasting one to one-and-a-half seconds. The downy often lives alongside a slightly larger look-alike, the hairy wood-pecker. The latter produces a more fevered pitch that is roughly two taps for every one of the downy's. The yellow-bellied sap-sucker of eastern North America and the red-naped sapsucker of the west begin their pounding in a rapid fashion, then quickly taper off into ever more widely spaced pairs of taps that make them sound a bit like they're running out of gas. Williamson's sapsuckers, northern flickers, and pileated, black-backed, and three-toed woodpeckers all have their own unique cadences.

You can listen to recordings of woodpeckers on any number of birding apps. Drum with your finger and try to keep up. Bring the animal's sound deeper into your psyche as you listen and move your finger to its beat. What we hear with our ears, process with our brains, and act upon with our muscles fires throughout our bodies. Our own movement sets in motion a cascade of internal electrical impulses and chemical responses that literally echo "'woodpecker" throughout our flesh and bones. After a couple of days or weeks of tapping along, you will likely start hearing woodpeckers *everywhere*.

The more we encounter and recognize animal signals around us, the wider the door opens onto learning what they communicate. When learning to play the saxophone in the fourth grade, I discovered the most remarkable thing: my world was overflowing with saxophones—they were in store windows, on television, in advertisements. People were playing them on street corners and in concert halls. Logically, saxophones hadn't suddenly been planted willy-nilly for my benefit; they had been there the whole time. I was experiencing what psychologists call the Baader-Meinhof phenomenon, or frequency illusion. This bias leads us to pick out more of the same based on past experiences. Thinking that playing the saxophone was cool, compounded by picking up the instrument and producing some squeaky notes, brought about a shift in my awareness. My impression was that saxophones were ubiquitous, but those brassy, bent instruments had been there all along. Not until I started to play did I take notice.

You may have encountered a similar noticing effect when researching and purchasing a product, such as a car. Selecting a vehicle of a particular make, model, and color can cause you to do a double take. How many times have you pulled on the door handle of the wrong vehicle in a parking lot? You start seeing "your car" all over town—on the highway, in driveways, and parked outside of school or your place of work. Your search image has become reoriented, even though all you've done is select a ride.

We recognize things when they interest us, and then we incorporate them into our lives. Humming along and tapping our feet to music is something we already do and is part of how we can name a song moments after it starts playing on the radio. If you are musically inclined, "imitate" a warbler by jotting down the notes of the animal's repertoire on a scale (some birdcalls fit into the eight-note scale of Western music and others fit well into the five-note scale common in Chinese music). If you dance or sketch, imitate the animal by moving your body or your pencil in a way that reflects the sounds you hear or the bounding gait of a jackrabbit or deer. My art students often find they create more satisfying renderings of a babbling brook by sketching with their eyes closed and moving their pencil along to the sounds than they do with their eyes open. Whatever you do, use as much of your own mind and anatomy as possible—and make duplicating nature's diversity fun. Joy and humor lead to improved memory.

Learning by doing fires our bodies and brains so that our neurological pathways can literally grow around the task. And it becomes easier and easier each time we try. Unfortunately, pushing the shutter release of a camera or the button of an audio recorder doesn't provide the same level of benefit as using more of ourselves, but it is still going to be better than not noticing at all.

Mimicking behavior not only makes things more memorable, but also inspires deeper understanding. As social creatures, we subconsciously merge with the others in our company; we may even begin to dress and act like each other. This "chameleon effect," augmented by mirror neurons in the brain, leads to us behaving similarly when we execute a task with someone else versus when we witness another person doing it. The drive to imitate is central to being human. It's part of empathizing with others; it's why we clutch the pillow during scary movies and wince when someone gets hit in the head with a Frisbee.

I learned of a class exercise carried out one December by a high school biology teacher in Saskatchewan that involved a powerful bit of role-playing. The teacher took the class out one cold, snowy day to a small natural area outside Saskatoon. The children, the teacher, and the parent helpers were all bundled up against the biting cold, and the teacher divided the participants up. Most were prey species. One was wolf, one was disease, and one—the teacher—was a hunter with a gun.

Wolf and disease could kill you only if they physically caught you; the hunter would kill you on sight. Once everyone knew their roles, the class split up. What happened next was fascinating. What would the prey species do to stay alive? The faster ones chose to give wolf and disease a run for their money, but when they did so, they had to stay out of sight of the hunter. Some simply hunkered down, but they would get hungry and need food or would get cold and need to warm up. Even in their hiding places, they might be spotted by wolf or disease and have no way of escape, but if they left, there was the constant terror of being spotted by the hunter. For all of those who participated, it was a visceral lesson in how vulnerable animals are when humans are about, and it showed how effective acting out these roles was when trying to understand another creature's point of view.

On applied and spiritual levels, there are other reasons why people copy what animals say and do. Evoking animals is a way to express reverence and gratitude, and to put food on the table. Indigenous cultures, such as the Lakota, imitate the buffalo or the prairie chicken with their ceremonial dances. Coast Salish performers shudder their forearms in an upright posture during their ceremonies to mimic an orca's quivering dorsal fin as it dives for deeper water. Members of Amazonian tribes act out certain sounds and events of a journey instead of only speaking descriptive words, as it puts the audience in greater contact with the substance of the story itself. Gwich'in hunters from central Alaska who find fresh bear signs might take cover and imitate the calls ravens use at a carcass to coax the bruin back so they can harvest it.

In art, zoomorphic representations of people are mixed with animals—think of the forty-thousand-year-old lion-headed person carved in mammoth ivory from a cave in Germany, or the rock paintings of creatures that are half-man, half-beast in Africa or the American Southwest. They show a veneration for these creatures, but perhaps also a desire to be like them and to harness their superpowers. When I am in my studio, I will sometimes step away from a sculpture I'm working on to perform my best stalking movements of a cougar or the trot of a deer. Embodying those creatures helps me figure out where an elbow or a foot might fall, and it elevates my understanding of those "others" in a global sense. How do they move, what do they see, what might they think or feel?

There are ethical considerations to be aware of when mimicking animals. As we strive to put ourselves back into direct relation to the natural world, it is important that we empathize with our wild kin. They are more than our esteemed guides and teachers; they can be our inspirations and role models. My grounding in imitating animal language initially came from those I hunted. My tune has changed a lot since then. I found that I could learn more about, and feel closer ties to, wildlife by keeping quiet and by creating non-disruptive reproductions of the animals I track, rather than by trying to engage directly in a dialogue with them. Just like when we pull the trigger on a gun, once we insert ourselves into the conversation, it either stops or it changes the equation entirely. From then on, all we see in their behavior are various reflections of our interruption.

I still cringe at the memory of walking around the Brown Farm in Virginia with friends. Someone began making a faux bird alarm call, known to birders as "pishing," to get a reaction from a hiding sparrow. Pishing commonly triggers birds to jump into view, or even approach the person producing the call, as they try to get a look at potential trouble. As the sparrow extricated itself from the dense undergrowth, a sharp-shinned hawk swooped down and snatched it. Where a living bird had been, only a puff of detached feathers drifting toward the ground remained. We were horrified. *We killed that bird.* There is a reason it's illegal to reproduce animal calls in protected spaces such as national parks.

Learning by doing lets you see further into details previously missed or taken for granted, like the woman in my class who got a glimpse into the goats' world when she realized the nanny had names for each of her children. Imitating animal sounds and movements can also give us a notion of what it might be like to be a member of another species. Experiment on your own time and in your own personal space. Remember that the reproductions you make will only be as good as the details your powers of observation have decoded. Allow yourself to look funny, sound ridiculous, and act oh, so immature. Move with the sounds, experiment with animal gestures, feel a reflection of the "other" in your own flesh and bones. Learning by doing is messy, but it can also be fun—and deeply connective.

PART 2

WHAT ANIMALS ARE TELLING YOU

− 9 −

YOU'RE
BEING WATCHED

UNLESS YOU ARE THE WINNER of the big raffle prize, being singled out in a crowd tends to trigger all sorts of unsettling feelings. Memories of uncomfortable moments at the class chalkboard or finger-pointing at a party come to mind. In other settings, our heart rates can quicken in an instant with a sideways glance from a love interest or the look from a child in a moment of discovery. Have you ever had a similar feeling when being watched?

Being plucked from the masses by an animal, however, is not something we generally give much thought to—that is, until one verbally assaults you. Our friend Shauna Baron had exactly this experience one autumn, with a bird. Shauna, a professional wildlife guide and former veterinary technician, was on her way back to her apartment after visiting a local café. Minding her own business, a coffee in one hand and a scone in the other, she snapped to attention after a raven dropped out of the sky and landed at her feet. The raven's motives were clear as it maneuvered with a sideways hop-step—it was there for a handout. Shauna addressed the bird in a conversational but incredulous tone, saying, "Oh, hello there. What do you think you're doing?"

In that exact instant, "it was like he exploded," Shauna recalls. The bird, now already in flight, landed in the closest tree at a height of about twenty feet and unleashed a verbal tirade. Shauna had never heard anything like this before. "It was giving a loud, harsh, repetitive scream." She felt the response was completely over-the-top. "It was as if I scared the bejesus out of him," she says.

Other ravens were drawn to the commotion, and they quickly joined the harassment campaign. Whatever slight had occurred, it was now the subject of conversation for an entire gang of corvid trash-talkers. The birds continued their scolding for the entirety of her walk back to her apartment. Did this raven know English? Did it pick up on something in the tone of Shauna's voice? Had the bird's guilty conscience been exposed—or was something else at work?

Shauna noticed a key detail before the episode came to a close—that raven had bling. Strapped to the bird's legs were a few colored bands, and on its back it wore a shiny black device. The latter was about the size of a Matchbox car with a couple of short antennas. Shauna had an epiphany—she and this bird had history.

Individuals of other species recognize their own kind. Even if you can't tell one gull from another, one bass from the next, their conspecifics certainly can, and they do it because they have to. It is essential to remember who kicked whose tail yesterday or last year, especially if it was yours. It is important to be able to recognize others of your own kind when you are deciding who might make a suitable mate or identifying your hungry, squawking kid in a crowd of hundreds if not thousands of others.

Domestic sheep register one another's facial features, and they remember up to fifty different individual faces for at least two years. Australian sea lions and northern fur seals remember other individuals' vocalizations for at least four years, particularly when the voice is their mother's. Bottlenose dolphins hold the record among nonhumans for the longest memory for old friends, and it may last twenty years or more. Elephants, primates such as chimpanzees and orangutans, and many marine mammals other than dolphins can easily pass the "who's who" test, as, it turns out, can some insects. Golden paper and hover wasps apparently use a combination of facial features and chemical cues to recognize hivemates.

Many species also recognize individuals of other species. Pet owners may notice that their companions often have different responses to different visitors. Your pony probably has a keen recollection of the veterinarian. Your dog might use a different bark for the delivery guy than she does for a neighbor. If you watch closely, you may also find that your wild friends make similar distinctions between individual humans. Spending a quiet morning on your surfboard or a peaceful afternoon in the hammock may not be the solitary activity you think it is. And if you've been naughty rather than nice, animals focus their skills when telling friend from foe. *You*, as it turns out, are being watched.

Even so-called lower organisms such as fish, insects, and mollusks can discern and recall specific humans from a lineup. Archerfish, honeybees, and octopuses—the latter doesn't even have a centralized nervous system—can ID your mug and remember it. As author Sy Montgomery notes in *The Soul of an Octopus*, giant Pacific octopuses can parse out people and are not above playing favorites. A technician at the New England Aquarium in Boston who got crosswise with an eight-armed cephalopod named Truman would get doused with a well-aimed jet of water from the animal's funnel whenever she entered the room.

The work of John Marzluff at the University of Washington provides a clear illustration of how well nonhumans can keep tabs on us. John and his colleagues have been conducting experiments with wild American crows on campus since the late 1990s; their discoveries have been featured on television shows, in a TED Talk, and in many popular publications. I find some of the more recent developments associated with this study quite fascinating.

Suspecting that American crows were able to differentiate individual humans using their facial features, John and his students were curious to see if tests could confirm it—ironically enough, so was the US government. The study, I would later learn, was funded in part through the US Department of Defense's Defense Advanced Research Projects Agency, aka DARPA, the DOD's "funder of weird and strange shit," as one scientist described it.

What interest might the US military have in studying the facial recognition skills of a common bird? We could speculate on all manner of reasons. Dolphins, sea lions, and beluga whales have been used in military programs by both the US and Russian governments as far back as the 1960s. In this case, according to John, the agency seemed most interested in finding missing persons.

From the beginning, the crow study was somewhat routine: capture some crows; weigh and measure them; fit them with a few colored leg bands; let them go; and follow up to see if they remembered their captors. That was it. From our perspective the protocols may seem innocuous, but from the birds' perspective, the whole thing likely resembled an alien abduction. The researchers hid their true identities by wearing a caveman mask with heavy brow ridges and cheek bones, a wide nose, and large projecting teeth. The idea was that biologists would later don the snaggle-toothed caveman mask to test how the crows would react. Could the birds identify their tormentor?

As researchers suspected, they could. The crows knew their abductor and reacted aggressively toward the caveman. The person wearing the caveman mask was subjected to raucous, scolding calls and dive-bombing flights by the birds. The crows essentially ignored everyone else, including the person wearing the control mask depicting former vice president Dick Cheney. The corvids' behavior made it clear that it was *this* face that had done them wrong.

The crows also recognized the mask when it was turned upside down—something we humans have a difficult time with. We are stymied by upside-down faces. One meme on social media shows the face of British singer-songwriter Adele turned upside down but with the eyes and mouth kept rightway up. The face looks fine to us despite the distortion. When the complete image is rotated 180 degrees, however, the Thatcherized image of Adele (so-called after former British prime minister Margaret Thatcher, whose image was first used for this purpose) becomes a hideous amalgam of misplaced features.

It was also surprising to see how long the crows remembered their captor. Following the one and only capture period, technicians would take the mask for a walk around campus once each ensuing year. As I write this, seventeen years after the experiment was performed, the crows still remember and harass the masked figure, even though only one of the original birds is still around. Caveman's nefarious ways have been codified into corvid lore. The deeds of this one foe have been shared, repeated, and passed down through the generations in a striking example of avian culture. What is even more interesting is that crows living up to three-quarters of a mile away from campus call out the masked figure. Word has spread.

When blackbirds dive-bomb a hawk or ravens harass an eagle, I can't help but think of John's results. Are they harassing a generic predator type or are they remembering this individual predator? And did the hecklers have a negative experience of their own or are they simply taking someone else's word for it?

I have seen only one instance of an eagle with a dead raven. In that case, it was a golden eagle tearing into one of these large corvids on the snowy floor of the Lamar Valley in Yellowstone. As the eagle removed the raven's feathers, breezes spread the charcoal plumes across the pristine snow, while a veritable tornado of fifty or sixty screaming ravens circled overhead. So important was this harassment campaign that the irate gang in black ignored a fresh wolf-killed elk only a half a mile away.

In Shauna's case, it was not her face that seemed to prompt her unusual reception, but her voice. The raven's odd reaction likely stemmed from her brief involvement with a new study on the ravens of Yellowstone led by Matthias Loretto, then of the Max Planck Institute in Germany, and John Marzluff.

With John back in Seattle, Matthias and I fired a net-launcher one snowy morning and captured twenty-two of these perennially hard-to-catch birds. In his smooth German accent, Matthias summed up the situation by saying, "Oh my goshe, Iy've nevah cote dis many!" How could we possibly mark, measure, and release all the birds before nightfall? A bit overwhelmed by what looked like a net full of bouncing, burnt popcorn, we needed backup. We sent out a few frantic texts, and Shauna answered the call.

Knowing about John's earlier work on crows, I had already covered my face with an orange handkerchief. I wore clothes I scarcely used and even donned an old pair of tattered running shoes. Furthermore, I did my best to keep my talking to a minimum. If there was any chance of pissing off the neighborhood ravens, I wasn't going to risk it.

Thinking I was a bit paranoid, the others carried on unmasked and in their street clothes. The use of a small cotton handling bag helped to limit the birds' movements and kept them calmly in the dark while shielding their captors from the powerful bites of those blood-letting beaks. What none of us bargained for was that even though they couldn't see us, the ravens were *listening*.

Shauna, excited to work with these clever birds, was full of questions. While leg bands and transmitters were being fitted, and wing, beak, and weight measurements were being taken, the ravens, it seems, were taking notes. Even though Shauna's scolding on the street was separated by a few days and a distance of two miles from the capture site, the raven's active mind was able to connect the dots.

If the raven had remembered Shauna's face, it would likely have exhibited the same response Matthias got a bit later from another marked bird—it cursed him from atop the roof of the apartment where he was staying. In Shauna's case, only after the raven heard her speak those nine words did it give that unforgettable retort. It may not be an exclusively human experience to remember the voices of your captors after being blindfolded and thrown in the trunk of a car. Although this sample size of one could be dismissed as anecdotal, I suspect that a repeat of John's study using sound instead of rubber masks might yield some interesting results.

Other animals parse out people by sight and sound in even greater detail than most of us would imagine. The wild African elephants of Amboseli National Park in Kenya can group humans by age, sex, and even tribal affiliation—and clearly know which ones mean danger. When articles of clothing are displayed and voices broadcast of agricultural Kamba people, the elephants seem unfazed. When Maasai sounds and belongings are substituted in, the response is striking. Specifically, it is the voices of adult Maasai men, not those of Maasai women or boys, that alarm the elephants. At the slightest hint of a Maasai man, adult elephants immediately assume defensive formations encircling their young. The elephants also flare their ears and trumpet loudly. Maasai men spear elephants as a rite of passage and demonstration of virility; the elephants' reaction is understandable.

Far from the wilds of Yellowstone or the African bush, similar things are likely going on in your area. How do animals in your neighborhood react to humans who harass them? Is there someone near you who shoots at the pesky starlings or who whisks the squirrels or pigeons away? Take note of how the different animals respond. How close are the troublemakers allowed to get before the wild observers get uneasy? How far up into the trees do they go to get away? What distance do they run before stopping and looking back?

You could also approach this exercise from the completely opposite direction, using yourself as the test subject. Try acting in a benevolent way. Be exceedingly kind and deferential to

all the creatures you meet. Give area nonhumans a wide berth.
Be hypersensitive to their every move, withdraw immediately
whenever they show signs of stress, talk to them gently, lovingly
even. Do they react differently to your woodland-sprite persona
than to your normal, everyday self?

If you are a good neighbor, word will undoubtedly spread.
Wild beings see us with great clarity, and when it comes to
matters of survival, it is not only ravens that are keeping score.
Kate and Adam Rice, a husband-and-wife photography team,
developed an affinity for a particular female grizzly bear in Yellowstone, starting in 2014. They spent so much time observing
and photographing this bear from a safe, respectful distance that
she came to see Adam and Kate as a benign, reassuring presence.
Bears who spend too much time munching on roadside vegetation—and close to unwitting tourists—are often hazed away by
park officials. Interestingly, the bears quickly learn to identify
government vehicles, and even individual drivers. In response
to seeing cars driven by gun-toting rangers, bears often maneu-
ver just beyond the effective range of cracker shells and rubber
bullets before those ranger vehicles ever come to a stop.

The Rices' caution and respect was not lost on the bears. On
occasions when their favorite female grizzly was presented with
a "bear jam"—a string of cars backed up on a park road for the
chance to see a wild grizzly bear—she would do something fascinating. When park rangers cordoned off a people- and car-free
zone to let the bear cross the road, she would pause, examine the
line of vehicles, and intentionally cross the road right in front of
the Rices' bumper. Kate and Adam estimate that she may have
done this more than twenty different times over the years that
they have known her. That particular van and those people

became equated with safety for her and her subsequent litters of cubs.

None of us are generic representatives of the human race. Wild creatures interact with us like they do with one another—as individuals. And even though animals and birds look superficially similar to us, we are interacting with *that* starling, *that* hawk, *that* horse, *that* pike, *that* cat, and *that* baker or candlestick maker. And all the while, we are being recognized, remembered, and discussed by our more-than-human cohabitants. How might your interactions with a nonhuman change after recognizing their individuality for yourself?

– 10 –

MAKING WAVES

TO EXPOSE MORE of what animals think of you, find a wild-ish spot close to home and sit quietly for an hour or two. This place should have a modest level of people traffic, such as one person passing by every hour or so—maybe a back trail at a local nature center or town park. Extra people accompanying you will add extra conversation, noise, and movement, so it would be a good idea to try this exercise solo. Avoid using paths right next to the visitor's center or parking lot, as animals there will likely be habituated to the steady stream of visitors. Explore your options beforehand and pick a trail that feels safe.

Select a place close to the trail where you can sit comfortably—about twenty to fifty yards away is ideal. Your chosen spot should have a reasonably open view and a nice backrest, such as a large rock or tree. Sit facing the direction that other travelers will likely approach from. As for gear, you won't need any, and in fact, if you have binoculars, you may want to leave them at home. Big-picture items will be the emphasis, rather than an up-close and personal view.

From here, it works like this: Get settled in your spot before the traffic begins. Morning is often best. Note that the wildlife activity may be a bit chaotic at first because of your arrival, or things may be completely silent. Know that this is part of the deal and remain patient. Let the forest or seashore dunes calm down around you for at least forty-five minutes or an hour. All that is required of you now is to look and listen. But look and listen for what?

First off, enjoy the peace and quiet—especially if getting out in nature is a relatively new experience for you. Absorb the sights and sounds of the location and bask in the feel of sunshine on your face. Let the calm settle your nerves as you open your senses. What I specifically urge you to look for is what a group of men camped in a forested setting near a crossroads in northern Virginia on the evening of May 2, 1863, observed, but failed to recognize. Their oversight cost many of them their lives and is likely something you have also been missing for your entire life, but will miss no more.

German immigrant soldiers of the Union XI Corps under the command of Oliver Otis Howard were settling down to their cook fires around dinnertime. While some men ate, others talked and played music. Sometime around 5:15 PM, a deer bounded wildly through the middle of camp. Soldiers laughed and pointed, and the deer was soon followed by additional forest animals. Rabbits, a flock of turkeys, and more deer soon burst through the crowded setting, and every last one of them was on the run. If an animal that is normally scared to death of entering a situation blindly bolts through it, then what's behind it must be far worse.

Moments later, twenty thousand charging Confederate soldiers under the command of Thomas "Stonewall" Jackson gave a blood-curdling rebel yell and descended on the Union encampment. The running deer and other animals flushed ahead of the advancing army should have been interpreted as a call to arms rather than a bit of comic relief. The fleeing Union soldiers were caught entirely off guard. They would later be dubbed "the Flying Dutchmen" and were only saved from complete calamity by the fall of night. The events would be remembered as part of the Battle of Chancellorsville of the American Civil War.

Once your wild-ish patch of ground has settled down, look for what happens as other humans start moving around. You, too, might spot a deer bounding at high speed as the first hiker comes down the trail. Alternately, but no less telling, the deer may walk a tad faster than normal but in a focused direction. The same deer may look over its shoulder or exhibit excessive tail-twitching. Keep monitoring. Scan for subtler things: diminutive birds, mammals, and other understated wild residents. Dramatic events will almost always be preceded by less overt behaviors. In the face of danger, flocks of small birds may take flight. Small mammals may pause and hover close to their burrows. Frogs may fall silent, and wetland fauna may move out of their element and into the forest. Had those at the Battle of Chancellorsville been paying attention, they would have detected many more precursory waves of unease in advance of the fleeing deer.

So, what exactly is happening here? A disturbance, whether an advancing infantry, a solo pedestrian, or a fox on the move, upends the peace in the immediate vicinity. The animals closest to the disturbance will be frightened and respond in a way that secures their own safety—like the way a squirrel darts up a tree or a rabbit bounds toward its warren. In addition to physically relocating, the town criers will voice their concern. Sharp chips or chatters replace softer, relaxed tones. Frogs, turtles, snakes, insects, bats, and fish also start behaving differently. At this stage of eavesdropping, it is not necessary that you know each species of animal. Simply notice what the animals are doing: Are they calmly feeding or scratching an itch, or are they frozen in place and all looking in the same direction?

No matter how nice a person you are, when you are out on the trail, the wildlings do not initially take your word for it. They judge your intentions based on the reactions of the first forest inhabitants you meet or, should I say, roust. First impressions are critical, and the size of the animal you first encounter is irrelevant. In nature's community of the safety-conscious, if you frighten the smallest among them, you have frightened *all* of them. Beyond the first ring of disturbance, the neighbors of those animals quickly take note. The chipmunk may drop its half-eaten seed and stand on a log for a better look. Droning insects may fall silent.

Everyone takes precautions as the threat advances. With a persistent threat, a wave of disturbance will travel farther and farther abroad. This pattern resembles the ripples after a pebble is thrown into a pond and is particularly distinct in the direction in which the threat is traveling. Some compare this phenomenon to the bow wave of a boat, such as the one dolphins gravitate to in the marine environment. These ripples in the calm can spread in all directions at speeds of over one hundred miles per hour. If the living waves can be recognized, they can be interpreted.

The first time I experienced "the wave" was back at the Brown Farm in Virginia; it was a revelation. Before classes one morning, I went out to an isolated corner of the farm to relax and enjoy the lively dawn chorus. I had sat out in the woods at dawn countless times as a hunter, but what I was about to see had escaped me entirely up until that moment. Seated below a ridgeline of mid-aged trees, amid a tangle of multiflora rose and honeysuckle bushes, I reveled in the sublime ambience of the space.

This subset of the farm, dubbed "Eden" by a fellow naturalist, lived up to its name. In addition to the broad diversity of birds found there, Eden was fertile, verdant, and blissfully quiet. I had been sitting for perhaps thirty or forty minutes when, all of a sudden, a flood of birds poured over the ridge and down into the vegetation around me. Chickadees, northern cardinals, a mockingbird, and titmice descended upon the bushes where I sat. A stunning white-eyed vireo landed within arm's reach to my left. It was heavenly. The birds almost seemed to be seeking me out and putting themselves on display. In truth, they weren't.

Minutes later, movement appeared along the ridge. Something furry and yellow morphed into a golden retriever. Following closely behind the off-leash dog was the owner. She appeared without noise or warning, but then it hit me—this flood of birds was a message. The dog, the woman, and the vireo—they were all connected. Everyone is party to the conversations of alarm—not just the snorting deer or chattering squirrel, but everyone.

When trouble shows up, word spreads fast. Erick Greene, a researcher from the University of Montana, has conducted field studies to look at nature's early warning systems. In one instance, Erick established a synchronized grid of microphones across a thousand-meter-by-thousand-meter plot of forest. Into one corner of this landscape, Erick piped in the prerecorded sounds of a chickadee cursing at a Cooper's hawk. The alarm bell's toll spread across the entire square kilometer within seconds. The sociable chatter of black-capped and mountain chickadees, red squirrels, and red-breasted nuthatches quickly morphed into jeering, agitated phrases of distress. The troubling meme had been passed from neighbor to neighbor throughout the entire woodland, and this frightful state persisted for over an hour. The primordial social media, the original Twitter, had done its job. When the neighborhood watch is in effect, predators often go hungry.

How often have you sat outdoors for an hour or more with no agenda, and just looked and listened? If you are like most people, the response is likely—never. This is why I insist, when teaching classes, that when we sit and observe, we do so for at least thirty or forty minutes, preferably longer. The result of not taking a seat is that you have never seen your area in anything other than a state of abject fear and disruption. Even some of the more accomplished birders I've met, ones who know every tiny chip note from every species in the book, often don't realize that those same notes are broadcasting news of danger—danger they themselves initiated. When we don't stay put long enough to become sensitized to the qualities of ease and stillness, the signposts of danger go directly over our heads. Having not heard any place in anything other than a state of alarm, it is no wonder we are missing what animals have to tell us.

On a day when Ishi, the last of California's native Yahi people, was teaching his physician—the author and legendary figure in the world of traditional archery Saxton Pope—to hunt with a bow and arrow, he paused and said, "The squirrel is scolding a fox." Pope could hear the squirrel's calls with his own ears but was incredulous about the claim. To Ishi's comment, Pope replied, "I don't believe you." After the two then took cover behind a rock and waited, a fox trundled by as if on cue—point taken. Not only had Ishi's powers of observation detected the wave of information passing over them, but he had deciphered its specific meaning.

Judy, one of my online-course students and a retired executive with the state psychological association living in Columbia, Maryland, discovered some of the same with "her" squirrels and foxes. As she watched from her elevated suburban deck, Judy found that the gray squirrels in her area make a sharper version of their bark alarm when there is a fox around. This differs from the call they make when they see the resident red-shouldered hawk. Judy adds that if the local birds join in the harassment campaign, she can collect further evidence about what the trouble might be.

While sipping coffee outside with her husband, Judy might say, "Do you hear that? The fox is down there." A second squirrel a hundred yards away might repeat the alarm of the first, to which Judy might announce, "Okay, now the fox has moved down there somewhere, and do you hear the crows and the wrens now? They're talking about it too."

If her husband raises an eyebrow, Judy might grab her binoculars and scan the area down toward the creek and confirm, as Ishi did, "Yup! There it is. The fox is *right there!*" The waves are specific, directional, and can be tagged to specific identities.

Ancient cultures have long told us that animals talk. The scientific community has only come around to these Indigenous ways of knowing in recent decades. Contemporary studies have confirmed that nuthatches, dwarf mongooses, scrub jays, yellow-bellied marmots, eastern chipmunks, and zenaida doves aren't just listening to members of their own tribes; they are listening to all of those present. The calls of hornbills, picas, woodpeckers, Carib grackles, and tufted titmice provide clues that they and many others take to heart. If the experiences of the Flying Dutchmen, Judy, and Ishi are any indication, we have much more to learn about wild conversations. The good news is, a master's degree or a PhD is not necessary. All you need to do is slow down, take a seat, and pay attention.

– II –

SCARIEST
OF THE SCARY

WHAT IS THE SCARIEST THING that animals talk about? Who makes the biggest splash? Is it sharks? Orcas? How about leopards, cobras, or wolverines? Unfortunately, the answer is commonly—us. Few things in nature cause such a stir as the presence of a human. Even before anatomically modern humans arrived on the scene, early hominids were hunting animals across large swaths of the planet. Except in places where wildlife populations remain entirely naive about humans—as in the early history of the Galápagos, Madagascar, New Zealand, and other island locations (Lewis and Clark even describe walking up to "fat and extremely gentle" wolves in eastern Montana and killing one with a spear)—our impacts are being felt.

Birds fly to the highest tree limbs, fish dart for deeper water, and creatures as large as whales and elephants gather in defensive groups when confronted by humans who might mean them harm. Hunting season clearly triggers deer, wild boars, and elk to hide, swim rivers, and run great distances to avoid their pursuers. Each one of us has likely witnessed some version of an unequivocal flight response. Our presence is felt up and down the food chain. As the apex predator, we not only hunt the prey species, we also hunt the hunters.

Even in a protected area with no hunting, Yellowstone's top predator—the grizzly bear—finds unexpected encounters with humans overpowering. Jenny and I briefly assisted with a research project conducted by Tyler Coleman for his PhD research on park grizzlies. Our task was to carry a GPS unit to record our travels in a remote section south of Yellowstone Lake. Tyler tracked our movements, along with those of other willing backcountry users, and overlaid them with those of GPS-collared bears. Few people, including ourselves, saw a bear, but our influence was being felt.

Tyler's study showed that the bears shifted their activity as soon as humans began using the trails in late spring and early summer. Grizzlies may have been hanging out in particular areas for some time, but as soon as hikers and backpackers showed up, things changed. Bears don't want to hang with people when given a choice. Some bears became active only at night, when the people were nestled in their sleeping bags. Other bears left the area entirely. In one instance, a GPS-tagged bear covered almost eleven miles in a single hour following a brush with hikers—that is one large, fast-moving ripple. Though he is careful to stress the associative nature of the data, Tyler's work seems to support the findings of previous investigators. Human impacts are undeniable.

Are we really that bad? With the way humanity operates these days, the answer is, unfortunately, yes. Species ranging from baboons to lions, weasels to wolves, birds of prey to predatory fish are all uneasy in our presence. Not the menacing sort? Odds are, someone who looks, smells, and sounds like you probably is. The term "the landscape of fear" is often invoked by behavioral ecologists to denote states of elevated vigilance by animals. No one is taking chances.

"Wait, wait, wait... hold on," you might be saying. "I'm an animal lover, I'm a vegan, I meditate. I have a ferret. I feed the birds. I go to sleep at night to the sound of David Attenborough narrating nature documentaries, I'm not *that* person!" All of this may be true, but that's not what your outward messaging campaign is projecting. You are still putting everyone on edge. The way you make eye contact, the tone of your voice, your body movements, even the volatile compounds coming out of your skin, can change the behavior of an animal in an instant. Myriad organisms feel discomfort around us, and this discomfort may be amplified when we look through the oversized "eyes" of our binoculars and camera lenses. Even if all you are doing is capturing a photograph, what you are saying is "I see you and I'm going to get you." And it goes even further. Humankind's influence may not require us to be present to spook wildlife.

Broadcast recordings of BBC nature-film narrators have been found to scare badgers in the UK. Recordings of National Public Radio hosts delivering the news or people reading poetry have sent wildlife in the US scattering. Playbacks of dogs and humans along the Oregon coast put raccoons on the move, and phantom highways and gas fields created with nothing but loudspeakers have reduced bird use in those areas by as much as 30 percent.

Loudspeakers belting out human voices in the Santa Cruz mountains of California were enough to drastically alter the movements of large carnivores such as mountain lions. The lions transitioned to routes farther from the speakers, and midsize hunters like skunks and opossums shifted to more nocturnal activity to find forage. The California study also showed that the effects of the piped-in sound reverberated all the way down the food chain to the local mouse population. The mice, however, instead of fleeing the human soundtracks, gravitated to them in an apparent effort to find safety and resources. You don't have to bulldoze a forest or tromp through the middle of an animal's turf to ruffle its fur or feathers. All you have to do is say "Boo."

The reason we don't see more animals is that they know we're coming long before we get there. I once scared a robin perched on the roof of our garage with a single step. My foot leaving the deck caused the robin to emit a harsh *cheep*. Its call was instantly repeated by another robin in the neighbor's willow tree about two-tenths of a mile away. In the span of a single second, news of my misstep was transmitted across the better part of a mile or more. Wild creatures using this high-speed messaging service understandably have a few minutes to react to someone's approach. How many minutes do they have? Naturalist and author Jon Young has found that a person moving at a walking pace is announced two full minutes before they actually arrive— you can almost set your watch to it.

Some animals automatically flee when humans are about; others hide in hopes you will walk right by. Cats are famous for doing this. A friend's house cat seems to enjoy hiding in plain sight when guests arrive and will not break from her post unless spoken to with direct eye contact. Some hikers or skiers have probably had the eerie experience of retracing their tracks only to find that they had been trod upon moments after they passed through. And if the prints belong to animals such as hyenas, lions, grizzly bears, or wolves, this can be somewhat unsettling. In North America, except in very rare instances, large carnivores are not tracking us to hunt and eat us. More often, animals learn of our approach through signals in the environment and step out of the way. Sometime later, that animal might loop back around to see "who goes there." Once their curiosity has been satisfied, they move on with their day.

See how you are being perceived by building an awareness of your surroundings. I try to remember to check my state of watchfulness at the front door. When I'm in a hurry, I'll often grab my things, turn the door handle, and burst out of the house, where I am met with a chaotic, frightened response. If there are birds on the feeder or if there is a deer standing in the lower yard, they beat a hasty retreat. If I open the door slowly and pause every once in a while as I make my way down the path, even for a second or two, the dynamic shifts. Try taking a look out the window before you exit; this can help a lot. Who can you see?

By pausing for a moment, having a look, then gently and slowly opening the door, you project a different message. A leisurely pace elicits a less acute reaction. You might stand outside the open door for yet another second or two, watching and listening. If you can manage just a few more seconds, take a step, but no more than two, then look and listen again. What do you find? Are the birds still sitting on the feeder? Has the cottontail rabbit ducked under the fence or is it still munching on grass with one eye directed your way?

In nature, news about us travels fast even if the transmission lines are not composed of copper wire or fiber optics. The conduits of nature's dialects lie in the minds, emotions, and flesh of the broadcasters. As we start to place our fingers on the quivering strings of this communal vibe, we too can feel it. Sift through the background noise, quiet your inner static, and become more aware of how the world sees us.

DECODING THE LANGUAGE OF ALARM

O NE HONK, CHIRP, OR HOWL is not the same as another. Listen carefully. Variations in the clarion trumpeting of a goose at the community duck pond or the yips of a ground squirrel can indicate details with startling specificity. In nature, small shifts in the soundscape can inform us of big things close at hand. Some calls differ in their duration, the number of notes or phrases, their frequency, intensity, or even the location from which they are broadcast. I will focus on alarm calls here because they offer a good opportunity for us to study cause and effect. Each permutation offers clues, but you will need to start paying extra-close attention from here on out. The devil, as well as the delights, will be in the details.

The sound hitting our ears is grating by any measure, and the outburst seems out of step with our calm surroundings. My friend Len and I pause. As the noise continues, he looks at me half smiling, half wincing behind his graying horseshoe mustache. We are being scolded—and loudly.

Beneath the vaulted ceiling of maples, red oaks, and linden trees, we could be anywhere. We are not in the forests of Canada or the North Woods of Maine, however. The two of us are meandering through a suburban woodland minutes from Len's house in Setauket, New York—a bedroom community to the Big Apple midway along Long Island's northern shore. This treed location is bordered on one side by busy roads, a CVS pharmacy, a strip mall with a nail salon, and a Whole Foods Market, and rimmed on the others by old neighborhoods and colonial homes dating back to the American Revolution. All of this is a mere sixty-mile drive from Wall Street and the financial district of Manhattan. This is not wilderness. Any illusion of isolation is erased by the din of road traffic, squeaking swing sets, and the shouts of children playing.

Above all other sounds, the persistent chuffing holds our attention. It's not a bird, it's not a plane, it's *Sciurus carolinensis*, a gray squirrel. This squirrel weighs only about a pound and is about the size of a roll of frozen cookie dough. Dressed in an ashen suit with a white belly and a tail about the same volume as its body, the squirrel continues its disquieting critique. It mixes the chuffing sound with high-pitched wheezes. One would think there is a Spanish Squirrel Inquisition in progress and that this tortured soul is nearing the end of its earthly days. The wave of sound, however, is intended to betray *our* presence.

The squirrel's call is telling everyone and everything that trouble is near. The sound is certainly meant to inform other squirrels, but it may also function as an emphatic "I see you and the jig is up" directed at troublemakers—in this case, Len and me. This siren of the woods is keeping the neighborhood safe, and for us, it's like trying to tiptoe around a sleeping household after the family's yappy dog has spotted you. The rodent's role as town crier has undoubtedly saved the lives of countless animals from human and wild hunters alike. Trying to be sneaky is futile once you've been spotted by one of these arboreal lookouts.

As we walk closer, the squirrel goes quiet as it enters stealth mode. Len and I find ourselves standing in a spacious void of silence. In the distance, a second squirrel picks up the charge, and then another about a hundred yards away. Chipmunks and even some birds join in the broadcast. We find ourselves in what feels like the calm eye of a building storm. Once we pass the squirrel's hiding spot, it renews its campaign.

When Len and I entered these woods, the squirrel greeted us with a low-frequency chuff call. In other situations, the same squirrel might switch to a high-pitched utterance that sounds a bit like a slow raking of fingernails across a chalkboard. Gray squirrels tend to use the husky chuff to out terrestrial predators and the fingernail call to call attention to fast-moving aerial hunters. Researchers are still disentangling the nuances of squirrel vocalizations, but the squirrels themselves do not appear to be confused—not when their lives are on the line.

When Len and I finally do get a look at the squirrel, it looks completely harmless; the rabble-rouser even looks cute. The rodent does a comical flourish with its tail and falls into another string of alarms as we walk away. And once it starts, it carries on and on . . . and on. This is exactly how it should be. The neighborhood watch is looking out for its own, but one sneaking question remains. Did the squirrel hear our car doors close way back at the beginning of our walk? Or did someone not as loud and conspicuous as the gray squirrel tip everyone else off? Did Len and I overlook chickadees while wrapped up in conversation? Was there a shouting robin or maybe a cussing cardinal who warned the squirrel?

An animal's behavior is dictated by its core needs and the constraints of the physical environment where it lives. The late Dutch biologist Nikolaas Tinbergen framed the matter of animal motives as a combination of "ultimate and proximate causes of behavior." A Nobel laureate and leader in the field of ethology, the study of animal behavior, Tinbergen held that the "ultimate" goal of any creature is to maximize its fitness. Fitness, per Darwin, has nothing to do with muscle shirts and workout rooms; it is all about living as long as possible and having lots of kids in the hope that those offspring live to produce children of their own.

While maximizing fitness is the animals' "ultimate" goal, Tinbergen held that animals' day-to-day actions are motivated by "proximate" factors such as eating to sate hunger, hiding to avoid danger, showing off to attract a mate, and drinking to slake thirst. Animals must satisfy these needs using what is within arm's, fin's, or wing's reach. Worldwide supply chains are not an option. The calorie economy is stingy and margins are thin. Songbirds in the far north may collect just enough food during daylight hours to last a single subzero night. Making the best use of precious resources is paramount.

One way to think of wildlife energy budgets is by comparing them to the dynamics of home heating. As the outside temperature starts to dip, adding heat to your house can help, but so can making better use of existing energy. Stuffing more insulation into the walls, donning a sweater, lowering your metabolism, or huddling with others can also improve the heating situation. Make no mistake, raising an alarm does have its costs, but when compared with other behaviors, chirping or bellowing to address predators or an adversary is a wise use of resources. Animals that run hither and yon at the first signs of trouble stretch already thin reserves. In other words, it pays to voice your alarm in the economy of life.

Going back to that Long Island woodlot, why would a squirrel care about a pair of wandering humans, or any other trouble for that matter? Wouldn't it be dangerous for a squirrel to spout off in the presence of trouble? The squirrel is a tree-dwelling rodent, after all—couldn't it just scamper up an oak or a poplar to a safe height and ignore mischief below? Why all the noise? Wouldn't the squirrel's commentary reveal its location and expose it to predators?

Life necessitates trade-offs. Speaking out can be hazardous to your health, but staying silent can also be risky. If a hawk or a fox showed up and no one spoke up, those hunters could easily pick off multiple prey animals and no one would notice. Having an active and vocal community watch saves lives. Having observant neighbors from multiple points of view makes for even better surveillance, even—or perhaps especially—if members of the neighborhood watch belong to many different species.

As Len and I saw, chipmunks and a few birds joined the harassment campaign initiated by the gray squirrel. Songbirds such as chickadees often travel in small groups composed of other chickadees, titmice, nuthatches, and brown creepers in winter. Extra eyes and ears attached to sympathetic voices give the group greater security and better feeding opportunities for all. There is even evidence that birds such as Carolina chickadees and tufted titmice fight more vigorously during territorial spats when they have a large, diverse audience watching. Think of it like a scuffle on the school playground, when having a circle of onlookers lets the combatants tussle without having to worry about the principal—or, in the songbirds' case, a Cooper's hawk—stepping in. Freed from having to constantly scan their surroundings, they can focus on landing punches instead. Eavesdroppers who listen to those raising the alarm have a choice of adding their voices to the community watch, continuing to feed, or going about whatever they choose to do knowing that the neighbors have their backs. The benefits of sharing and receiving information, it seems, keep everyone's options open.

The squirrel, proximately speaking, called Len and me out either because it saw us or heard us walking along or because it was reacting to an alarm raised by other woodland creatures. We were the stimulus, and the squirrel gave a response. The squirrel only halted its harangue when Len and I got close enough to see it. Prudent gossips keep self-preservation in mind. They stay attentive as the situation develops, then melt into the shadows so that they might live another day, find mates, and multiply. Now that you know what to listen for, I'm sure you'll quickly identify the gossips in your neighborhood and see how they are keeping an eye on you and broadcasting your movements to the wider community.

Returning to the wolf and coyote episode outlined in the introduction to this book, we saw how something as simple as a coyote barking on a hilltop can deepen our understanding of what is going on around us. In *The Expression of the Emotions in Man and Animals*, Darwin asserted that wild dogs do not bark, save the coyote of North America, but this is not true. Wolves, coyotes, and even the purported non-barking dingoes of Australia give a shout out to trouble, just not very often. Barks are what your dog does when someone unexpected shows up. Barking is a warning. If you were to keep company with dingoes, coyotes, or wolves every day for an entire year, you would hear a bona fide bark on only a handful of occasions—far too infrequent for most observers to make note of it, much less decode the meaning.

Real barking in these wild canids typically only happens in the presence of imminent threats, such as enemy wolves, domestic dogs, grizzly bears, or humans. That's about it. The list of irksome characters is quite short. After I imitated the alarm bark of a gray wolf for a class some years ago, a student who worked at a captive wildlife facility exclaimed, "That's exactly what the wolves do when the veterinarian shows up!"

Apparently, the facility's wolves were not fans of veterinary visits any more than most household pets are. The wolves took great umbrage at the vet's arrival and greeted him with a volley of scathing bark-howls. Sometime after learning about this unique call, the young woman was at work when the wolves suddenly erupted into a raucous fit of "bad veterinarian" barking. She thought it was odd, because the veterinarian wasn't supposed to come until the end of the week. Scheduled or not, the animal doctor's truck was in the parking lot and in view of the wolf enclosure. The vet had come earlier than expected and the wolves were not going to let the event go unheralded.

As far as coyotes go, there are few things that raise their hackles like the presence of a wolf. Along with humans and disease, wolves are a leading cause of death for coyotes in the wild, and as soon as one arrives, news spreads. When a coyote alarm-barks, it does so in a repeated, staccato manner, and for a sustained period.

If you think you hear an alarm bark once or twice, the coyote could be raising an alarm or it could be communicating something else. But if the ruckus goes on for fifteen minutes or more, then you'd best have a look around. This suggestion also applies to other animal alarms from squirrels and birds to frogs. If your nervous system hasn't already automatically done it for you, tune in to the consistency and persistence of the alarm call you are tracking. Investigate the corner of the town park or local woodlot where the chattering is coming from. An owl might be perched neatly out of view, camouflaged amid the boughs of an evergreen tree. Without the ruckus, you never would have seen it.

Specifically speaking, a coyote's distinctive upturn in pitch near the end of its bark-howl note is a telling clue that the big dogs are in town. While coyotes seem to have variations on the alarm bark for other threats, to me, this upturn is the wolf-specific signature embedded in *Canis latrans*'s most disquieting call—and it is the exact coyote vocalization I heard that morning in Mammoth Hot Springs.

After the reintroduction of wolves to Yellowstone Park in the 1990s, the coyotes immediately began using their bark-howl to out wolves, much as that gray squirrel in New York outed my friend and me as we strolled under the trees. Before the wolves were released from their acclimation pens, armed guards protecting them noted "coyotes with sore throats" yelling at the wolves. Jennifer Sheldon and Robert Crabtree were the first to recognize this "yip-howl," as they called it, as unique. Even after a wolfless interval of seventy years and many generations of carefree coyotes, they hadn't forgotten their word for *wolf*.

This atavistic sound of warning, like the alarm calls of many species, seems to be innate rather than learned. And here is another delicious detail of the smaller canines' alarm: they do not make this bark-howl when they smell a wolf, nor do they make it when they hear them—they do this only when they *see* a wolf. The call is so predictable that it tells me there's a wolf around even when I'm two or three miles away across the expanse of the upper Yellowstone River valley.

If you hear a barking coyote, the next thing to do is find the individual animal that is making all that noise. You might have some time to do this, because if the wolf sticks around, the bark-howls can be repeated for as long as forty-five minutes. If you can see the coyote, look in the direction it's looking. Commonly, within two or three hundred yards, you will find the wolf. Once you learn to recognize this distinctive vocalization, it's hard to ignore. All you have to do from there to pinpoint the pack or individual wolf is pick up a pair of binoculars or a spotting scope.

Although millions of people have seen wild wolves in Yellowstone, and despite it being one of the most reliable places on the planet to catch a glimpse of them, many millions more have missed their opportunity. Visitors miss their chance for no other reason than they don't know these wild conversations exist. Most visitors never know that the wolf they most want to see is around the next bend in the road or over an adjacent hill. If these travelers only knew that a subscription-free, no-email-required wildlife communication network is sitting unused at their fingertips, they might have had a very different trip.

Coyotes verbally harass cougars too, and may have a tailor-made version of the bark-howl for these big cats. The cat call, as I have witnessed it, is a mouthier "husky" howl embedded in the barking coyote's cry. Although my personal dataset has taken many years to amass and is still quite small, I've confirmed this lion-specific epithet on several occasions by direct observations or by looking for big cat tracks in the snow after the event has taken place. Many biologists I have spoken with have yet to recognize this howl for cougars; however, several wildlife film-makers who regularly follow these cats have noted this call and a few have captured it on camera.

The coyote alarm bark for a feline is heard less frequently than even the alarm howl for wolves. Owing to lower density across vast territories and stealthy habits, lions are perennially hard to pin down. YouTube, ironically enough, has added some data points to the evolving picture. Phone-toting travelers in places like Banff National Park in Canada or Yosemite National Park, California, as well as homeowners who have installed trail cameras in their backyards, have documented these unique warnings and uploaded the evidence to the video-sharing website. I have compiled a playlist of coyote alarms on my YouTube channel. If you live in an area where there are mountain lions and pay attention to what the coyotes are saying, you stand a much better chance of seeing one of these elusive cats.

Like so many other animals, coyotes tend to monitor dangerous situations from a safe distance. *Canis latrans* will often maintain a two- or three-hundred-yard buffer between themselves and wolves, but remember, this is a response to visual stimuli. In forests or grasslands where visibility may be limited, the distance between the two will be correspondingly less. How much less? That depends on your local conditions.

As the threat begins to move, the town criers often shadow it. This conveniently betrays not only the location, but the direction, and sometimes even the speed, at which the threat is moving. When I heard the change in that coyote's position that morning in the park, I knew the Canyon Wolf Pack was on its way *downstream* along the Gardner River and roughly how fast the pack was traveling. Listening to the shifting location of the coyote was key to disentangling those details.

A duck alarming at an otter as it follows the aquatic predator could yield this same kind of information. Experienced law enforcement officers will sometimes employ this shadowing behavior when dealing with a suspect leaving the scene of a crime. Will the suspect reveal their accomplices? There is nothing magical about reading these signals; it's all about making better observations. If you are someone who is involved with neighborhood monitoring, working as a biological field technician, or helping with a bit of citizen science, nature's signal system can be priceless. If you combine your observational skills with modern technology, you will place yourself in the upper echelon of wildlife investigators, and even if achieving this status is not one of your goals, knowing such things provides endless satisfaction.

For example, some years ago, Yellowstone Park researchers told us that a pack of wolves was in a nearby valley. We positioned ourselves on a hilltop overlooking that valley and trained our eyes and spotting scopes in the direction the researchers said the pack's radio signals were coming from, but to no avail. What people often don't realize is that radio signals can bounce wildly under certain conditions. When researchers use radio telemetry, they focus on the loudest signal. The idea is that, as the operator swings the directional antenna and narrows down the arc of the loudest beeping tone, the antenna will end up pointing in the direction of the collared animal. What happens in mountainous terrain, and even over water bodies, is that the echo of the signal can be as loud as, if not louder than, the direct signal. The result is that many researchers, including me, have been embarrassed to find that they are off by as much as 180 degrees.

After spending about an hour looking in the wrong direction, our group suddenly heard a coyote bark-howling from behind us. We had driven right past a fresh wolf kill and eleven satiated wolves less than a hundred yards from the road, only to stand on a hill and look in the opposite direction. Had we not tuned in to the coyote's broadcast, we might have never known the wolves were there.

– 13 –

SPEAKING CHICKADEE-ESE

TO UNDERSTAND HOW wild warnings might unfold in a regular back-yard setting, one need only look to more of the work done by Erick Greene from the University of Montana. It can be challenging to draw meaningful conclusions about scary encounters when all you have is a small number of observations—such as my suspicions about a coyote alarm for cougars mentioned in the last chapter. To solve the problem of predator encounters being episodic and unpredictable, Erick invoked a bit of science-driven sleight of hand, or rather, bird, to help, and conjured up an unlikely assistant—meet *roboraptor*.

A local taxidermist and electronics whiz turned Dr. Franken-stein concocted a series of avian monsters for Erick's work. Each creation combines the skin of a predatory bird such as a pygmy owl or a Cooper's hawk, some Styrofoam, wires, motors, and a garage door opener. The fake hunters come with moving heads and tails, and each is set on a concealed perch behind a fake hol-low log. The log shields the roboraptor from view until precisely the right moment.

Why would Erick adopt this approach in the first place? In the late 1970s, Dorothy Cheney and Robert Seyfarth conducted landmark research proving that nonhumans—vervet monkeys, in this case—used discrete, word-like sounds to communicate. The vervets of Amboseli National Park in Kenya make three distinctly different calls in the presence of a snake, an eagle, or a leopard. When Cheney and Seyfarth broadcast a recording of a vervet's alarm for "martial eagle" from a hidden loudspeaker, the monkeys glanced up into the sky and immediately hid in dense bushes. If an alarm for leopard was played, they leaped into the trees and looked down. If the warning for python was used, the monkeys stood on their hind legs, the better to scan the tall grass around them. The recordings and playbacks helped tease meaning out of a jumble of undecipherable sounds.

Other studies have used doppelgänger predators, such as a stuffed jackal on wheels pulled through a colony of meerkats or a fake snake made from the skin of a dead python and presented to a group of chimpanzees. Much of this research has been done in far-off places with exotic species, but Erick Greene was now bringing it to North American homeowners' backyards.

In an online presentation he gave during the 2020 pandemic, Erick noted that this line of work called for some of the cushiest observation blinds he had ever used. Sitting comfortably behind insulated walls and large picture windows, and plied with hot drinks by supportive homeowners, Erick was living the high life for a field biologist. When it came time to get to work, Erick would hit the garage door opener and trigger the big reveal.

As the log slowly lowered, it exposed a heart-stopping predator directly in the midst of the placid neighborhood setting. Area birds responded with what can only be described as feathered fireworks. Quiet conviviality was shattered by an intense display of sound and movement. The worst fears of each tiny bird had just been realized—a serial killer had dropped directly into their midst, weapons brandished. Imagine someone slipping into your bedroom in the dark of night and releasing a *Tyrannosaurus rex* or a pack of velociraptors ready to consume you the instant you pull back the covers.

Erick's flock of frightening birds allows him to make the presence of imminent danger both predictable and recordable. When comparing the results with pre- and post-scare audio and visuals, it's easy to see that when a killer crops up, the number, volume, and intensity of birdcalls dramatically increase. Each chickadee, and soon each nuthatch, sparrow, and kinglet, is bouncing around in bursts of frantic flight. They move up and down, side to side, forward and back. Each passerine's parrying is carefully executed to maintain a safe distance from the specter of sharp beaks and talons.

Every time the roboraptor's head spins or tail jiggles, it elicits wave after wave of intensified alarms from the amassing flock. This mob scene builds further as more sparrows, nuthatches, and chickadees, as well as robins, creepers, and even squirrels, join the fray. Harassment of the roboraptor quickly becomes a neighborhood affair.

The black-capped chickadee's onomatopoeic name is derived from a vocalization it often uses as an alarm call. In Erick's work, he has found that the chickadee's alarm call not only pinpoints the location of trouble, but may also encode the threat risk and the physical qualities of the predators themselves.

To clarify before we go any further, the *chickadee* call is not the chickadee's song. Songs are often complex, repeated, and seasonal, and are largely used for the purpose of reproduction—attracting mates, carving out a place for the next generation, and keeping other males at bay. The chickadee's song is a whistled *seee saww*. Occasionally, a third syllable will be thrown in, making the song sound a bit like *cheeeze burrr-ger*. Calls generally address everything else. They are usually shorter and less intricate than songs and can be heard throughout the year; they are commonly used as an alarm. More *dees* equal more danger. Erick's roboraptors are helping unravel the strands of this story in a backyard close to you.

Kathryn Sieving from the University of Florida studies the language of tufted titmice. The *dee* call, as it turns out, is a stock-in-trade vocalization for an entire family of birds known as parids. The Paridae family includes not only chickadees and titmice, but several species of tits found across Europe and Asia. Katie's work on the meaning behind these calls began with a simple question asked by a student: "Why doesn't pishing work in the tropics?" *Pishing* is using an imitation bird alarm call like the one that was used on that fateful day on the Brown Farm in Virginia. The student, son of a professional bird-watcher and guide, was familiar with the technique of mimicking the alarm call of parids to expose all manner of hidden songbirds. But here's the catch: parids are found only in the Northern Hemisphere. Birds in the tropics don't recognize the northern equivalent of an ambulance siren going off, because they have never heard one before and therefore don't react.

Katie and her students have decoded many parid vocalizations and continue to amass new discoveries. On one occasion, Katie was taking a morning walk along a path with one of her students when she heard an unusually low frequency *dee, dee,*

dee call from a titmouse. It stopped her in her tracks. This variation was something new to Katie and she was at a loss for what it meant. Turning around to look, she and her companion spotted the titmouse in question, but then also saw what it was talking about—a bobcat slinking through a low, wet area just off the trail.

When I'm out walking with Hobbes, our Labrador, with all his retriever antics, chickadees often elicit strings of four to six *dee* notes after the *chicka* part. A dog is a ground threat, and though bothersome, Hobbes is of limited danger to these birds. When I leave Hobbes at home and walk alone—moving quietly, going slowly, and paying attention to my surroundings—I am met with single- or double-noted *chickadeedees*. One morning, Hobbes and I heard a cascade of twelve or more *dees* fired off in rapid succession. We were not being particularly menacing, and I certainly wasn't wielding a shotgun aimed at mowing down our black-capped buddies. As it turns out, they weren't talking about us at all. In the mid-distance, I picked out movement from the corner of my vision. Out of an adjoining stand of aspens, a slashing horizontal blue-gray streak cut through the vertical tree trunks, revealing what they were *really* talking about—a Cooper's hawk.

I had another interesting encounter with chickadee alarms one day when I had gone out to a sit spot close to my home. It was December 6 and the outside temperature hovered around 20 degrees Fahrenheit. My goal was to take a seat somewhere pleasant and passively observe what events unfolded. This might seem like a setup for a bit of mild torture—early in the morning, dark, stationary, and cold—but to limit any prolonged discomfort, I planned to keep the outing to exactly sixty minutes, no more, no less.

For my hour-long vigil, I chose a post known to our family as Sitting Rock. The rock was disgorged from a glacier's icy innards around fifteen thousand years ago and broke in two. The two granite fragments later separated and came to rest in the shape of a large lithic sofa. Smattered in blotchy gray and black lichen embroidery, the rock is further accented with a few crusty patches of lichen in pastel green and mustard yellow. The "seat" of the sofa rests a few dozen yards from the thready waters of a creek and offers an excellent observation post. When I sit here, I feel intimately connected to the landscape as I gaze over riparian alders and willows, across sagebrush hills, and beyond to distant mountains. That day, I set my watch, turned off my phone, and turned on my senses.

The first fifteen minutes elapsed with little of note. A few distant ravens croaked and a magpie or two showed up, but nothing more. I began to wonder if this was a productive use of my time. Doing my best to extinguish distracting thoughts, I reminded myself to pay attention. No sooner had I silenced the inner dialogue than I heard a voice along the creek. *Chick-a-dee-dee.*

The chickadee's call resides solidly within a frequency range that my ringing ears can hear. More importantly, this bird was saying something, something very specific—but what? Having remained entirely silent for the preceding quarter of an hour, the chickadee spoke at that very moment, not before, and as I would discover, not afterward. The bird only uttered a single call with two *dees* at the end. The *chicka* part was squeakily garbled and a bit difficult to make out. Could it be a low-grade alarm, or was it something entirely different? My mind began to spin with the possibilities. Again, I had to halt the endless loop of thoughts and return to the moment.

A light northwest breeze jostled the grasses, and the creek gurgled beneath a layer of ice. The sun lifted above the eastern horizon and a few wispy clouds, looking as if they had been applied with a sponge, formed high above. But I heard no more chickadee calls. I concluded that it must have been nothing—a misfire, perhaps? Then, as if by magic, a stunning silver-coated coyote appeared not twenty feet away. The appearance of the coyote was a shock, and I felt a rush of adrenaline spill through my veins.

I didn't move a muscle, but on the inside, I was bouncing around like a jar filled with overcaffeinated grasshoppers. "This is so cool. It is *so* close!" Breezes licked the ebony hairs along the coyote's back, lifting and flexing them ever so slightly in the direction of its head. With so little space between us, I could clearly see the muted orange tint on the back of its ears. One of those ears also had a small nick along its outer edge.

Did the lighter fur and the presence of the injury suggest that this was an older individual? The coyote looked left momentarily, then faced the rising sun. The direction of the breeze was perfect. We were cross-wind from each other and the coyote did not smell my bouquet of lingering breakfast, hand soap, and other distinctly human odors. Despite the openness of the landscape, it didn't pick my shape out either. This is another reason why I put my back up against Sitting Rock; it broke up my silhouette.

I was being super careful not to look directly at the coyote either, one of the other lessons I learned during my years as a hunter. A great number of people might whirl around, face the coyote, point a finger, and exclaim out loud, "Look, a coyote!" Don't do this. Always look by moving your eyes first; then and only then, if the animal is at ease and, ideally, looks in the other direction, turn your head its way. Direct eye contact is one of the greatest affronts to animals in the wild; it's a sign that the animal's cover has been blown and it is now receiving a direct challenge from you, the observer. Keeping my eyes low, I shot only fleeting glances at the wily coyote.

The craziest thing was that I never saw this coyote coming. It was as if someone had dropped it into an animation reel—in one frame there was nothing, and in the next, there it was. Later, I realized that the coyote had followed a shallow draw from the creek bed up to where I was sitting. The declivity concealed its approach until it was right next to me. In more developed settings, this sort of thing can happen via an irrigation ditch, sewer line, underpass, or road culvert. The chickadee, as it turned out, was not simply blowing off steam; it was sounding an alarm, albeit a low-level one. And now I understood why. The coyote was not hunting. It was traveling in a calm, relaxed manner and was probably headed to the neighbors' bone pile—a place where they dump the leftovers from hunting season's butchering activity.

Before moving on, the coyote calmly looked back over its shoulder, a subtle but telling clue that it may have had a companion. I instinctively looked in that direction, but saw nothing. Bolstered by the encounter, I reclined back into my stone couch, but I was a bit too quick to pat myself on the back. A minute later, a Townsend's solitaire gave its characteristic single-note call about two hundred yards to the north. This vocalization can be used for a few things, but like the chickadee's *chickadee*-ing, it can also be used as an alarm. An American robin soon barked using a sharp *peek* and a red squirrel let loose with a volley of descending, chattering notes. Was this a minor territorial squabble between squirrels or did this also mean trouble?

Cumulatively, the robin, the squirrel, and the solitaire added to my suspicions that other factors were in play. Might there have been a second coyote or, at the least, another disturber of the peace in the vicinity? Were they all gossiping about something out of view to me? As it turned out, this is exactly what was happening. I would have to wait for the full picture to take shape.

A Crow tribal Elder once asked a photographer friend if he would like to know where to find wolves and bears. The photographer waited eagerly for the intel, expecting the older man to draw a map or give a detailed description of likely places. Instead, the Elder imparted an unexpected sort of wisdom. "You have to listen to the birds," he said. "The birds are like our old women. They gossip about everything." This bit of advice was presented in the most matter-of-fact way, as if the Elder was telling the photographer that he'd get run over if he crossed a busy street without looking both ways. The simplicity of that statement conveyed its eternal truth. Animals *are* exchanging information, and my experience at Sitting Rock reinforced this fact.

After my morning viewing session was over, young George and I took a walk. Our destination was the patch of forest north of Sitting Rock. This is where the solitaire, squirrel, and robin calls had come from. Unfortunately, there was no snow to help us find tracks—that is, until young George and I wove our way beneath some dense junipers. There, on a lingering patch of snow no larger than a modest floor rug, was a string of fresh eastbound coyote tracks. The calls had been given for good reason: there *had* been a second coyote.

I suspected that one of the two coyotes, a female or an older individual, perhaps, may have detected something out of place. With potential trouble in the creek corridor, the coyote seems to have changed course and given the entire creek bottom a wide berth. Maybe it caught my scent or noticed an unusual hush enveloping the creek itself. Regardless of the reason, that second individual turned widely to the north, walked through the stand of trees, and presumably rejoined its companion after passing my location.

The birds, the squirrel, and the observed coyote gave me a much wider understanding of the local landscape. People often turn to apps such as Nextdoor or local Facebook pages to connect with neighbors about an upcoming bake sale or to monitor conditions during security alerts or natural disasters; the non-human network was doing much of the same.

There is one more level of alarm that would be helpful to discuss before we move on. Code red for many birds is conveyed with *seet* calls similar to the fingernail call employed by that gray squirrel in New York. When a predator is moving fast and close, birds of all shapes and sizes will emit their own version of high-pitched *seets*. Unfortunately, with my hearing loss, I no longer hear these vocalizations well, but I can see the response to them.

Seet calls cause area avifauna to freeze. It is as if birds are hit with a life-or-death pause button. Bird-language practitioner Dan Gardoqui refers to these statuesque creatures as "birdsicles." When a *seet* alarm is sounded, no bird moves. Whether they are snapping up grain at the feeder, gathering nest materials, or preening in a thicket, they all stop what they are doing and hold still. The pause may last for several protracted seconds and in a bubble of complete silence. A friend once played a recording of a chickadee *seet* alarm at his bird feeder, which caused a white-breasted nuthatch to go rigid and stay frozen for a full minute and five seconds. The nuthatch broke the pose only after a nearby chickadee gave the all-clear call.

The high frequency *seet* alarm is both short in duration and notoriously difficult to locate. It's like the songbirds' secret frequency. Many hawks and owls do not hear much above 7,000 Hz, and hence the 8,000–11,000 Hz *seet* alarms fly above their radar. You could compare this to those teenagers who sometimes use high-pitched ringtones on their mobile phones to avoid detection by their adult teachers and parents, whose ears are less sensitive to these frequencies.

When it comes to close-in, fast-moving trouble, birds seem to share a global vernacular. In his travels, Erick Greene has played recordings of these short, high-pitched *seet* calls and found that they, unlike parid *dee* calls, are recognized on almost every continent. In Australia and Europe, and across North, South, and Central America, birds that hear this sound instantly freeze, look around, then dive for cover. In an amazing feat of evolutionary trickery, some caterpillars have even evolved the ability to generate a *seet* sound as a way to protect themselves from being eaten by birds. Saying *seet*, as it turns out, projects a message that is globally read and clearly understood.

Depending on where you live, tallying the number of *dee*s, frozen birds, or minutes of incessant barking by a coyote in your midst can be key to finding interesting events you might not otherwise have seen. Pick one common call you hear local animals give and begin observing how it modulates with time and context. Try scribbling down some notes so you can refer back to them later, maybe imitate the calls, or record them using the video or audio functions on your phone. Whatever methods you use, having ways to compare things over time can help you understand a lot more than you would by merely capturing a single snapshot.

As you carry out your investigations, you may discover that certain species respond differently to the same "wolf" in your backyard. What flusters the robins today may not be the same as tomorrow. Why is this? You may find that you end up with far more questions than answers, and in the end, this is a great reason for another outing. Departures from commonly heard calls can set the stage for recognizing an amazing array of details in wildlife commentaries.

– 14 –

VARIATIONS
ON A THEME

G RASPING A DOMINANT THEME of a simple animal call can be a useful entry point into the expansive topic of animal language. Start by picking a single species or, better yet, one vocalization from a single species. You will probably hear your wild neighbors saying many of the same things over and over. Grab a hold of these frequent expressions, dissect them, look for the circumstances and timing of their use. Once you feel like you have a handle on that most common call, you can step deeper into the unknown.

My own efforts to learn from ravens began with this simple, incremental approach a few decades ago, and the exciting journey continues to this day. Identifying one three-note phrase enabled me to segue into knowing the location of an owl-killed rabbit, deciphering the details of territorial spats, forecasting the approach of a golden eagle, and a lot more. Fans of classical music will know that variations on a symphonic theme can conjure all sorts of mental imagery, not to mention hints about the inner landscape of the composer. Before diving into the raven variations, some lessons from a ground-dwelling rodent from the American southwest might help.

Con Slobodchikoff, now a professor emeritus at Northern Arizona University, spent part of his career studying the conversations of prairie dogs. Con's results provided some startling insights into language-like communication in nonhumans. Whether nonhumans can even have something we can label as language (rather than the term "referential vocalizations," used by researchers) is still a matter of debate; Con himself was quite skeptical until the data began mounting. Where did Con begin? He started with a simple theme: prairie dogs signal differently when threats arrive by land or by air. Starting with this simple concept, Con ultimately found that the chatter of Gunnison's prairie dogs bears an uncanny resemblance to human grammar and phonemes—the latter being discrete sounds that we organize into countless meanings, such as the way the letters *b*, *a*, and *t* might create the words *bat* and *tab*.

Con Slobodchikoff didn't set out to study animal language. His initial questions were focused on the social dynamics of prairie dogs. Only after a previous advisor, Peter Marler (who also worked with Dorothy Cheney and Robert Seyfarth on the vervet monkey studies), prompted Con to see if prairie dogs might have different calls for aerial versus ground predators did he take a look. This kind of vocal specificity had recently been shown for ground squirrels in California, and now Con was finding it in prairie dogs too.

At a certain point, Con began to wonder if, beyond categories of threats, the prairie dogs might be describing features of the predators themselves. To the neophyte ear, prairie dog sounds could easily be confused with the chirping of birds, and many required computer analysis to disentangle the details. With time, however, some interesting patterns emerged. The data showed that not only did the animals have calls that varied in reference to air or ground attack, but they had additional calls for coyotes versus dogs, and for hawks, badgers, and humans, some of which could be parsed out by the unaided human ear.

Not only did the prairie dogs have a wide variety of calls, but they also seldom, if ever, misspoke. In one instance, the researchers saw a German shepherd entering the prairie dog colony. The rodents alarmed as though it was a coyote, and the researchers were delighted to think they had finally found an instance of a mistake. Understandably, a German shepherd could be mixed up with a coyote at least once in a while, right? When the canine approached closer, the field crew realized that the mistake was on them—it *was* a coyote.

As they delved deeper into the alarms that prairie dogs used for humans, Con and his students began changing the clothing, body types, and behavior of human test subjects. Although the details of each call became harder to discern with the unaided ear, the computer analysis continued to show that the calls fit the different categories used in the experimental trials. For instance, there was a unique call used in the presence of a tall person wearing a green shirt—versus blue or yellow—walking slowly through the colony. This differed from calls for a short person with a green shirt moving quickly.

The prairie dogs were exhibiting a form of grammar in which snippets of a short, squeaky sound functioned as nouns (*dog* versus *human*), adjectives (*fast* versus *slow*), and verbs (such as *flying*, *walking*, or *running*), and all of these could be used in different combinations to suit different situations. Con also found that the prairie dogs created new combinations for novel objects. In one experiment, Con's team introduced a painted plywood oval to the colony and the animals came up with a new "word" for it.

Con's thinking began to coalesce around something language-like in a nonhuman, but many of his colleagues were not so sure. Presented with evidence for higher communication skills in prairie dogs, some scientific publications mandated that Con change the human-oriented linguistic terms when describing prairie dogs' calls. Words like *nouns, verbs,* and *adjectives* were off-limits, as they were considered to be human-only constructs. Even though Con adhered to sound scientific protocols and made use of the best available technology and analytic tools, his interpretations made some who hewed to linguistic orthodoxy uncomfortable.

Con still has his critics, but the weight of current scholarship seems to be moving further in his direction. Of the features defining human language—following lists such as the one created in the 1960s by linguist Charles Hockett—only one, recursion (the nesting of ideas within other ideas, such as "Chris said that the birds would start migrating, but only after Amy pointed out that the weather must change"), is still considered exclusively human. This, too, however, is being drawn into question by ongoing studies. As interesting as this sort of research and the resulting theoretical debates are, we must ask: Can someone off the street—without the aid of recording devices and analysis software—decipher detailed meaning in animal communication? The answer to that is most certainly yes—as the ravens taught me.

Throaty croaks rang out through the limpid morning air. A raven and its mate were perched on the timber arch framing the corral gate beside our modest cabin. Jenny and I had recently relocated from Virginia to work for the Yellowstone Association— Yellowstone Park's educational nonprofit partner. Home became the Lamar Buffalo Ranch in the heart of the park's northern range. As Jenny and I settled in, the ravens became a constant presence.

Most days began with the same feathered scenario. The ravens and I were far from best friends. If anything, we could be described as coolly distant cohabitants. Each morning, as I set out to do chores such as stacking firewood, pulling non-native plants, and waiting for participants to arrive for a wildlife watching tour, there were the ravens. When I went for a hike, there were the ravens. Whenever and wherever I went fishing, wildlife watching, or sketching—there was Raven.

Even with my degrees in wildlife biology and despite becoming an astute bird-watcher, I had to confess that I didn't really know this bird. I knew its name in English and in Latin, but I did not grasp the ravens' ways. I did not comprehend their likes, dislikes, familial habits, taste in roadkill, or anything else. The birder in me could readily identify a raven in flight by its longer wings and wedge-shaped tail, and by its tendency to soar in the wind instead of constantly flapping the way crows do. I could distinguish the raven by any number of throaty *caws*, croaks, and *quarks*, so different from the clear, higher-pitched calls of the American crow. In time, I could even distinguish the raven's tracks in the snow or mud from those left by magpies, crows, or eagles. I was a walking field guide, but like many field guides, I was long on names and identifying marks, but lean on other content.

Even though I had recorded the ravens' presence in my notebooks for years, collecting observations the way a philatelist amasses stamps or an oenophile collects bottles of merlot or

sauvignon blanc, I had not pushed through to the level of con-
noisseurship these birds deserved. Living with them now in my
new home in the valley, I desired—no, I *craved*—a sense of belong-
ing to this new landscape. Ever since leaving New York and then
Virginia, I had felt untethered. It was as if I was floating about,
drifting this way and that, looking down on the real world from
high above. I had a strong urge to be a part of a place again, to
have community, and I hoped these winged residents could help
me. At first, the ravens held tight to their hard-earned wisdom,
but gradually they began to let me in.

The most common raven call I heard from that cabin in Yellowstone was a triplet of *caws*. I had no idea what this common
theme meant. The insistent trio of notes seemed to be repeated
in an unending, almost monotonous loop. I should mention that
during this time I was staying away from any and all raven literature. I did not want to be influenced by what others had come to
see. In time, I would consult outside opinions, but for now I was
still shouldering a bit of academia-induced PTSD, having just
finished my graduate work. For the time being, it was important
to me to only look and listen.

The three notes sank deeper into my consciousness; they
percolated there, yet did not reveal much. It was not until the
ravens changed their tune one day that my world shifted. As
the two birds on the corral arch did their usual thing, another
raven entered their airspace. One of the corral birds let out a new
sound that I hadn't heard before, then gave chase.

The trick, as I discovered, was to take this one theme—*caw,
caw, caw*—and look for instances where it changed. It was that
simple. Then it dawned on me: this was the resident pair who
ruled this section of the valley. The three-note call was their way

of saying, "Mine, mine, mine," or "Stay out, stay out, stay out." The triple caw acted as an acoustic No Trespassing sign. This is how other songbirds employ their melodies, and as ravens are the largest songbirds in the world, it makes sense that they would have a tune too, although a raven's song is unlikely to inspire composers the way the nightingale, quail, and cuckoo moved Beethoven when he was composing his Sixth Symphony.

The nature of the raven's song began to morph as my neural networks were firing in the background. The pace of the calls would sometimes quicken and come with extra notes. I might hear four, five, or eight *caws* in a row. Gradually, I sensed that the faster cadence and extra *caws* were expressing an elevated sense of possessiveness. When a new roadkill squirrel showed up or some tourists broke out a bag of snacks in a parking lot, the commentary in raven-ese shifted from "Mine, mine, mine" to "Mine-Mine-Mine-Mine-Mine-Mine!" Whereas more *dees* in the chickadee's call equates to greater danger, more of the same for ravens seemed to translate into heightened rapacity.

Then there was the time when I witnessed the use of two *caws* instead of the usual three. A group of ravens that had climbed high into the sky above a cliffside late one morning began playing on the windy updrafts like otters of the air. Periodically, a raven would drop toward the earth and then spin in midair, performing a perfect barrel roll, before swooping upward. During the final phase of their free fall, the birds would croon with two *caws*, as if saying, "See that!?" I shared this with a friend from Northern California and he was delighted to report the next day that he saw the exact same sequence unfold at his house.

In Chopin's nocturnes and Mozart's multiple takes on "Twinkle, Twinkle, Little Star," there is an implicit understanding that small modifications in sound can create broad shifts in expression. Adjustments to tempo, timbre, instrumentation, pitch, volume—pianissimo to fortissimo—alter our experience and our mood. The Adagio for Strings by American composer Samuel Barber hangs over a room like a lead weight. The Adagio has been described as the saddest piece of music ever written—it was broadcast on the day of President John F. Kennedy's funeral. Bach, Barber, the Beatles, Benny Goodman, Led Zeppelin, Del McCoury, John Lee Hooker, Lady Gaga, and Dr. Dre have all used the same palette of notes, but it is the manner in which they used those notes, and, equally important, the space between them, that affects us so.

Edging further into the repertoire of these corvid composers, there was another call that crept into my awareness. More erratic and jumbled than other utterances, this ravenism was a hodgepodge of unmatched notes with inconsistent spaces between them. The emphasis was different too. Inflected notes went up and down, and they were repeated once, twice, or not at all. In contrast to the territorial call, this longer phrase resembled *KAH, Ka-ka, kuh–Kah, ko-kuh, ka, Caw.*

Before I could even see the bird making this call, my ears told me that the speaker was on the move. It turned out that there was not one bird involved, but two. One raven was pursuing the other. Their jerky flight pattern made them look like two boxers in loose-fitting black trench coats dodging punches. Wings went up and bodies went down; both birds darted from side to side as the trailing bird attempted to push the other downward. This "chase call," as it is termed, is part of how ravens deal with trespassers. This is exactly the call I heard when one of the corral pair pursued the interloper. The chase call became another anchor point for variations on yet another theme.

Wild turkeys are another species that can reveal depths of meaning if you are willing to immerse yourself in their world. Florida naturalist and author Joe Hutto is someone who is willing to do this. The wild turkeys he studied had a range of expression extending far beyond the dozen or so calls accepted in scientific circles. For example, if the birds found a rattlesnake or a copperhead, they used sounds distinct from those they used for other types of snakes. Before his time with the turkeys, Joe might see only one or two venomous snakes each year in that area of Florida. Through the birds, he came to understand what they had found, and after that, he might see one of those snakes each day.

Joe tells me in a phone conversation that even before they were fully grown, the turkeys would spread out in a "skirmish line" perhaps thirty yards across to forage. "I would hear a vocalization and I would know with 100 percent certainty that it was [for] a diamondback rattlesnake and not a gray rat snake. There was absolutely no question that there's a rattler, and over there a gray rat snake, or a black racer." When I press him for specifics about what made one call special over another, Joe answers, "I wouldn't necessarily be able to discern the difference in the call, but I could tell the difference in meaning, if that makes any sense."

Joe likens the communications of wild turkeys to the layers of an immense onion. Rather than peeling this onion from the outside in, this journey progressed from the inside out. Layer by transparent layer, his understanding of the turkeys' world expanded. Variations in the turkeys' calls extended far beyond the classification of snakes. Hatchlings appeared to be born botanists, mammalogists, ornithologists, and entomologists. They instinctively knew which species were harmful and which were edible. As he delved deeper into their vocabulary, Joe would test himself by waiting until he heard a turkey call out something he interpreted as "Box turtle." Joe would then walk over to see if what they had found really was a box turtle. Gradually, his turkey vocabulary expanded.

Even though Joe understands some "words," he realizes there is so much more he does not know. The more time he spent with the birds, the more Joe realized that the communication between the turkeys extends far beyond one sound equating to one thing and a different one corresponding to another. There is an ethereal beauty that contains many mysteries. "It goes further out into the vastness of that onion," Joe says. "It goes out to infinity. I came away with my mind being boggled and the realization that this is so complicated that I just don't have a clue as to what is actually going on, and to think I would have the arrogance to suggest that I could possibly understand it."

I feel those sentiments acutely when I listen to the ravens. But like Joe, I discovered that the birds gave me enough remedial language-arts lessons that I could elevate my awareness of what was going on in their world severalfold. With time, those glossy black Beethovens came to show me much more. Ravens told me of traveling hikers, bedded wolves, distant carcasses in shaded ravines, the direction a bear might be moving, where human hunters might have found success, and the location of an otter's fishy dinner.

Variations on the raven-on-raven chase call would soon clue me in to the presence of other species—especially eagles. By paying attention to the ravens' incessant *ka-ka-ka-ka-ka-ka-*

ka-KA-ka-KA-ka calls over the years, I have discovered many golden eagles that I would not have seen otherwise. Biologist David Haines of the Yellowstone Bird Project has also known about the raven's golden-eagle chase call for decades and has used it reliably to find eagles in several settings around North America.

I also feel ravens have a different chase call for bald eagles. Since golden eagles are far more lethal to ravens than the fish and waterfowl–focused bald eagles, it is logical that the ravens would differentiate between the two. The raven chase call for a bald eagle sounds to me a bit like that for a golden, but with overtones of what they say when giving other ravens a tongue-lashing. Additional chase call variations exist—for red-tailed, Swainson's, and ferruginous hawks, and for peregrine and prairie falcons—but these are not yet cemented in my mind.

The messages are there. We can find them by giving ourselves the gift of time and attention. As I mentioned at the beginning of this chapter, it pays to take time with a common creature close to home and identify its three or four most common calls. Out of those, take maybe the top one or two and start honing in on the smallest of variations. Although passive sound-capture with a recording device can be useful, active exploration—such as taking a short walk to see why the animal changes its tune, the way Joe did when learning turkey-speak—will lead you to more nuanced understandings. Sketch a visual representation of the sound or gesture, apply phonetics to write out a memorable phrase fitting the tune, jot down some notes on a scale, or maybe simply whistle the sounds you hear.

The constellation of sounds coming out of your local animals will be myriad and often quite puzzling. Don't let this act as a barrier; see it as part of the game. I will sometimes hear single croaks emitted by a lone raven at mid altitude. I'm still baffled. Froggy-sounding chuckles from treetop ravens, or pairs in flight saying things like *glug-glug-glug* or *coo-coo-coo, kah-kah-kah*, leave my head spinning. Water drops, snaps, clicks, and gurgles are all in the mix, along with a zinger that sounds like *wahhh-wahhh, wahh-wahh*, which confuses me to this day.

To help keep track of it all, I finally broke down and made a raven vocabulary list on my cell phone. The act of making the list helped me differentiate the vocalizations better. It's helpful for me with "my" ravens, though I realize it touches on none of the diverse regional accents and dialects of ravens in other locations. Ravens in Maine, for instance, speak differently from ravens in Montana and Wyoming; they sound a bit higher-pitched overall, to my ears. California ravens are a bit more nasal—"Dude, check out the dead squid on that dude's surfboard!" Those from Virginia, Alaska, and northern Canada vary too, and I am told the trend applies throughout ravenland in Europe and Asia as well.

Just last year, a student of mine from the Netherlands sent me a video with the inquiring caption: "Is this the raven word for wolf?" He had recorded the clip just before an excited cyclist told him about a large gray wolf that had slipped into the undergrowth in that spot. The footage panned across a bank of deciduous trees with a raven making a pithy, edgeless, and upturned *whhaaaaAA*. I had never heard anything like it before. Is this the Dutch raven word for wolf? I have no idea. Could this be a new entry in a northern European raven lexicon? Maybe. When the student sent the video to a contact who monitors a wolf pack in Germany, he replied, "That's what they say here when there's a wolf around."

I had occasion to ask professor emeritus and leading expert on ravens Bernd Heinrich how many different meanings might be encoded in raven vocabulary. Both Bernd and one of his protégés and a master of corvid studies himself, John Marzluff, gave me the same answer: "We don't know." The onion keeps expanding. As vast as raven-speak is, finding a couple of common, repeated calls proved to be an accessible foothold that allowed me to enter their world, and they have shown me facets of my home that I never would have seen without them.

− 15 −

HOWLING MYSTERIES

REGARDLESS OF HOW LONG we watch, no matter how many experts we consult, there will always be mysteries. One striking example comes from our friend Shauna (scolded by the raven earlier) and a group of captive gray wolves.

Some years ago, Shauna received a rousing reception, a queen's welcome, when she returned to the Mission Wolf sanctuary outside Westcliffe, Colorado. Her visit came on the heels of an extended hiatus when she had been traveling abroad and working in the Rocky Mountain west. Ebullient howls resonated from the wolf enclosures as Shauna approached the facility. Interestingly, the gleeful yips, barks, and sonorous tones started not when she pulled into the parking lot, but when she was still a mile away. No one knew Shauna was coming—except the wolves.

Thirteen miles from the nearest paved road, Mission Wolf lies within a broad expanse of sagebrush, aspen, ponderosa, and pinyon pine in view of the Sangre de Cristo Mountains. Some of the wolves and wolf-dog hybrids residing there are rescues from well-intentioned but ill-prepared homes. Others come from film productions that acquired the animals for a shoot, then jettisoned them after they "wrapped." The sanctuary offers these animals a second chance at life.

Mission Wolf cofounder Kent Weber had brought Shauna on several years earlier to help with regular chores, caring for the wolves, and conducting educational programs for the public. Before long, Shauna was an integral member of the sanctuary's operation. She lived at the facility on and off for the next four years. Strong bonds developed between Shauna and many of the canine residents that later staff would have trouble getting anywhere near.

Curiously, on the day of her return, the wolves seemed to anticipate her arrival. When they heard the wolves howling, the staff suspected something special was about to happen. The wolves had exhibited this unusual greeting behavior before Shauna's time at Mission Wolf, and by now it was a bit of an inside joke.

· "Who's coming back to visit this time?" someone might ask. "The wolves would start howling and carrying on like the meat truck was coming down the driveway," Shauna would say. The meat truck is an old white pickup used to collect roadkill, deceased cattle, and horse carcasses from local ranches to feed the packs, and every once in a while the wolves tuned up their meat truck howl when that vehicle was nowhere to be found.

It didn't seem to matter how long the absence of said person had been either—whether it was hours, months, or, in Shauna's case, three years. Wolves, it seems, don't forget a friend. Kent told me that individual wolves have remembered specific people after an absence of ten or twelve years. The longevity of their recall appeared to be limited only by the age of the wolf itself. The sanctuary gets a lot of visitors. Dozens to hundreds of cars file in and out each day in the summertime, and the wolves scarcely make a peep; this is part of what makes these greetings so interesting.

Taking a survey of a wolf's senses may offer some background on how and why the wolves do some of the things they do. We have already seen how the super-sniffers of dogs can detect deceased animals through river ice and packed earth, but could the wolves be using their sense of smell to predict the arrival of old friends? If wolves' knack for identifying people in a crowd is any indication, they surely can. During Shauna's time at Mission Wolf, she had an eye-opening experience with one particular ambassador wolf named Rami.

Rami, a small black-coated female wolf weighing seventy or eighty pounds, was fifteen months old when she and Shauna met. Born in a litter at Mission Wolf after a botched vasectomy, Rami was both relaxed and confident around people—a model ambassador wolf. When Rami was four, the Mission Wolf team began taking her on the road to local venues and, eventually, to the coast of Maine. On one of these cross-country tours, Shauna, unbeknownst to her colleagues, set up a test for Rami at a stop in Vermont. The results of Shauna's experiment are awe-inspiring.

Before the program started, audience members filled every one of the two hundred or so seats in a large outdoor tent. Attendance was high enough that a standing crowd of ten additional rows gathered at the back. As Shauna addressed the gathering from the stage, Kent Weber led Rami into the crowd for introductions. Rami began making the rounds in her usual fashion, but she soon started to pull at the leash a bit harder than normal. As they entered the center aisle between the seats, Rami started dragging Kent into the standing crowd at the back.

Rami forced herself between people, going seven or eight rows deep. Shauna couldn't see exactly what was happening, but within a second or two, Rami had reversed course. The wolf then pulled Kent back toward Shauna on the event stage. As Shauna puts it, "Rami started going nuts and doing this excited, squealing sound, and her whole body was wiggling with excitement."

Shauna felt Rami was saying something along the lines of, "Shauna, Shauna, come here, come here *quick!* Follow me. There's someone you need to see!" Shauna had secretly invited her mother, sister, and stepfather to join the event from their home in upstate New York. Shauna's invitation came with explicit instructions: "You can't meet me at the touring bus or greet me. Don't do anything that might draw attention to yourselves—just make sure you are in the crowd, *and be ready!* If Rami finds you, grab her cheeks, because when she greets you, she could give you a bloody nose with her forceful kiss."

Through a riot of human odors, amid the noise and distractions of all the people, Rami's nose led her straight to Shauna's mother—and she gave her the promised kiss, sans nosebleed, thank goodness. Rami had never met Shauna's family before that day. Shauna would later note that her sister garnered similar attention from Rami, "but only about half as much." Rami completely ignored her stepdad.

The wolf was not just picking out a specific smell in her environment the way an animal might sniff a distant carcass. Rami was detecting relationships. Rami replicated this act of relative-finding in Ohio, with the family of one of the other handlers—and with precisely the same outcome. With a poke of her probing nose, Rami also succeeded in pinpointing breast cancer in two women on that trip, just as some dogs are trained to do. In

Rami's case, one of the women knew she had cancer; the other did not. The latter put Mission Wolf in her will.

During her time at Mission Wolf, Shauna and her fellow volunteers quickly learned that each wolf saw the other animals, as well as the humans, as distinct individuals. They saw each of them as unique and treated them accordingly. The signals the wolves used ranged from overt to exquisitely subtle and applied as much to human-wolf relationships as they did to wolf-on-wolf ones. "You had to clear your mind and go into the pens without baggage," Shauna tells me. If you didn't do that, the wolves might single you out for attention, and not always in a good way. "It was almost as if they knew something was up without you saying it or displaying it."

This doesn't mean, however, that working with captive wolves inevitably leads to confrontations. Mission Wolf was intentionally designed to be anything but combative. In the 1980s, when Weber cofounded the sanctuary, the prevailing wisdom was that you needed to be forceful, domineering, and alpha-like with captive wolves. This mindset went against Kent's personal philosophy, and he chose a different path. He gave wolves the same respect he gave humans. Many observers—including other professional animal handlers—comment on how relaxed and even playful Mission Wolf wolves are.

Kent took some of his inspiration for working with the wolves from natural horsemanship or "horse whisperer" techniques. He doesn't use food as a reward and doesn't ask anything of the wolves unless they show a talent for something. Then, as Weber says, "they might get a job." Weber notes that earlier research suggested wolves had difficulty reading human cues. Although wolves that have been socialized and trained from their first days of life can follow a person's signals the way a dog can, training a wolf generally takes more effort. When a person points, a dog is predisposed to look in the direction indicated; a wolf raised with little or no contact with humans simply looks at your hand. After spending decades in close company with gray wolves, Kent knows wolves react differently—wolves don't follow your hand, they follow your *eyes*. Kent uses his eyes and body to subtly direct his ambassador wolves, working with the wolves' innate tendencies, rather than forcing his will upon them.

Kent has enabled upward of a million people to have a direct personal experience with a wolf. Much of the power within those encounters, he feels, is grounded in the nature of the creature itself. "Wolves have a way of looking right into you," he tells me. That sense of being seen clearly by someone or something has an undeniable effect on people. Youth with autism and those with histories of abuse light up. Adults, such as veterans with PTSD, find sincere, life-changing solace in lobo's company. Kent tells me that many troubled souls get more benefit from a day or a week with the wolves than they would from a three-month intensive wilderness-therapy experience or years of counseling.

Is this part of the "wolf medicine" that Indigenous cultures talk about? "They look through the trauma and the injury, and see *you*," says Kent. A standard greeting with a wolf involves sniffing noses. The full monty requires you to smile wide so they can lick your mouth and teeth. If this makes you squirm, you might want to mind your manners. Denying a wolf this simple multisensory gesture of goodwill would be like refusing to say hi or not accepting a handshake. After an initial greeting of a group, the wolf will often select one or two individuals to spend a little extra time with.

"Without fail," Kent says, "the wolf will pick out someone that fits into one of only a few categories. They will gravitate to the bully, the scapegoat, the valedictorian, or class clown, or, invariably, someone with trauma." Kent has seen this selection process repeatedly unfold in groups as small as a dozen and as large as several hundred, and even at public venues with crowds numbering in the thousands. How can a wild animal survey so many people in a matter of seconds? And how do they isolate individuals that psychologists and school counselors have been trying to reach for months or years? Some have wondered if wolves are able to see auras or something like them.

The way the wolves behave seems to suggest that we have a clinical diagnosis pinned to our shirts, or are wearing one of those big sandwich-board placards announcing our mental and emotional status for the world to see. Feeling neglected and despondent? A wolf may feel it. If someone is projecting "I can't shake that instance of assault" or "I've lost the one person in the world that truly cares about me" or "My mind is still on a tour of duty, even though my body is back in the civilian world," the wolf seems to know. Kent reinforces this by saying, "Teachers would come back to me and ask, 'How did the wolf know to go to that one kid? He's the one that lost his mom yesterday' or 'She's the one that had something terrible happen in her family.'"

Though these things may not be apparent to human eyes, they seem to be rimmed in neon lights for *Canis lupus*. The human body has been found to emit ultra-weak light, which varies with our circadian rhythms. It might also fluctuate with our mental and emotional states, and *if* wolves are able to detect it, could they be seeing us wearing our hearts on our sleeves? Odds are, it has more to do with things like levels of eye contact,

body posture, and, in particular, smell. Over time, Kent has concluded that wolves are attracted to people who are aware of their
surroundings, and they are unsettled by those stuck in tunnel
vision. But we might be remiss to think it ends there. So, what
else are wolves assessing?

We've known for some time that dogs of all stripes can track
down their caretakers when separated. They can detect cancer
and the onset of rapid swings in blood pressure or blood sugar,
and even identify the ashes of cremated people, such as those
who perished in the deadly wildfires of California. Research has
shown that dogs can be trained to sniff out indicators of stress
and PTSD as well. Volatile organic compounds such as cortisol,
and perhaps as many as one hundred other chemicals, may be
emanating from our skin, and they can say a lot about what's
going on inside us. Given that wolves and dogs have nearly identical olfactory systems, it's not a stretch to think that both can
smell many of the same things. But whatever Kent's ambassador
wolves are doing, they are doing it with *zero* training.

Carl Safina, in his book *Beyond Words: What Animals Think and
Feel*, features an account of wild dolphins encircling a research
vessel, but refusing to come within fifty feet of the boat. The crew
discovered that one of their own had died belowdecks. The dolphins proceeded to flank the boat—not wave-ride in the front as
usual—and seemed to escort it back to the docks. Did their echolocation tell them that a heart had stopped beating? It's hard
to say.

According to Kent, a raucous greeting howl often starts at
Mission Wolf when the familiar person parks their car and steps
out into the open. Seeing the individual triggers the howl. In
Shauna's case, however, she was in her vehicle a mile away with

the windows up. Could it be that the wolves were able to hear Shauna coming back, as opposed to seeing or smelling her? I have watched wolves communicating with other wolves across large distances. Humans with normal hearing can often detect a wolf howling from two-and-a-half or three miles away. We have documented wolves directly conversing at distances of six or more miles. Dave Mech, one of the world's leading authorities on the species, has found that in open terrain, wolves can hear other wolves as far as ten miles away.

Kent says his staff will often hear the wolves initiate a greeting howl when he is still about seven miles away. At that distance, Kent's car is barely visible on the horizon. A wolf might connect a distinctive engine sound or rattling suspension with the impending arrival of a specific person, but Shauna came back in a different vehicle. If the wolves could not see Shauna or know her vehicle by what their ears told them, how could they make the connection?

This phenomenon of knowing from a distance is a puzzling one. In a case with elephants and the late conservationist Lawrence Anthony, the pachyderms knew when he was home and when he was away. As their savior and closest human contact, Lawrence was literally the lifeline for this group of rogue wild elephants destined to be euthanized. Lawrence brought the problematic group to his wild animal reserve, Thula Thula, in Zululand, South Africa, where they flourished.

In his memoir, Lawrence writes of being away on business in Durban and returning to find the elephants right there to greet him. The first time it happened, he chalked it up to coincidence. "But," he writes, "it happened again after the next trip and the next." Somehow the elephants knew when he was home and away, and then, as he puts it, "it got . . . well, spooky. I was in an airport in Johannesburg and missed my flight home. Back at Thula Thula, 400 miles away, the herd was on their way up to the house when, as I was later told, they suddenly halted and turned around and retreated into the bush. We later worked out that this happened exactly at the time I missed my flight. The next day they were back at the house as I arrived." When Anthony passed away in 2012, the elephants walked for twelve hours to reach his home and remained there silently for two days.

Walling ourselves off from contact with wild nature deprives us of connection to the wider world and its inherent mysteries. If we can't see it, hear it, or smell it, we tend to think something doesn't exist. Yet, when we get to know the "other," the world becomes a bit broader, more diverse, and far more inspiring. Our furred, scaled, and feathered kin have much to impart, and through them we see that there is more to this world than meets the eye.

PART 3

GOING DEEPER INTO PLACE

– 16 –

1 + 1 = 1,000

TO BE HONEST, I was not quite sure how a group of seventy people could learn much of anything about animal language together. I was accustomed to studying wildlife dialogues on my own, and the thought of joining a class this size seemed like the antithesis of being stealthy and inconspicuous. Walking in the woods with even a single trusted friend can generate too much noise and movement, but walking with seventy strangers? The educator in me wanted to see how this might work; and besides, I had a long overdue thank-you to deliver.

My journey to participate in this program, entitled Bird Language Intensive, began long before I hopped on a flight to San Jose, California. The seed of this trip had been planted twenty-one years earlier, following my experience with the birds, the hiker, and the golden retriever at the Brown Farm in Virginia. A friend, knowing of my interest in birds, tracking, and all things natural history, shared something quite new.

"I've got something you might be interested in," he said. "Give me some cassette tapes when you get a chance and I'll make you a copy."

The only tapes I had were my trusted *Real Turkeys*, volumes 1–6, on the vocabulary of wild turkeys by Lovett Williams. These recordings were the soundtrack to my multiyear love affair with turkey-speak and that trip to the state calling contest in Albany. In the absence of any blank cassettes, I handed over my *Real Turkeys* tapes on faith, not knowing what I was going to get in return. Some days later, the bootlegged tapes arrived, bundled together with a green rubber band, wrapped in paper, and labeled with black handwriting that read "Concentric rings."

The full title of the audio series was *Advanced Bird Language: Reading the Concentric Rings of Nature* by a naturalist named Jon Young. I instantly recognized the merits of the recordings. Here was someone who, like me, had struggled with the perplexing issue of how animals such as foxes were so ninja-like in their ability to know when humans were in the woods and then stay completely out of sight and out of mind. Better yet, he had been developing ways to teach these skills to others. Fast-forward two decades and I arrive at what looks like a mini Scout jamboree not far from Año Nuevo State Park, just north of Monterey Bay on the coast of California.

People from all walks of life, including teachers and massage therapists, attorneys and college students, are setting up tents and getting oriented. Many of these folks fall solidly within Jon's existing fan club. We have committed to five days of camping and dining on locally sourced fare while participating in group lectures, discussions, and field activities, all on the subject of bird language. Besides coming from a variety of generations and backgrounds, the program participants span a range of experience, from rank beginners to studied practitioners. I'm curious how this might work once we start watching the birds.

Standing beneath a large white event tent, Jon welcomes everyone and lays the groundwork for the program; he is a slim, affable sort with a thin salt-and-pepper goatee and graying hair extending beyond the back of his blue knit cap. As I expected from the recordings and the online lectures I watched before the workshop, he seems easygoing and eminently approachable. Jon's limitless curiosity is matched by his gift as a storyteller.

Jon begins by posing a question: "Why would we want to know bird language in the first place?"

He shares that much of what we dedicate our attention to relates to the severity of the consequences if we don't. Let the time run out on a street meter and you'll get a ticket. Ignore the electric bill and the lights go out. Have you checked the contents of your refrigerator or investment portfolio recently? If a bird says the hyenas are out hunting at night, you keep your kids inside. If life or death literally hangs on the thready notes of a bird or rodent sound, you learn to listen. Jon reinforces this point by sharing a compelling story from his travels.

The San people of the Kalahari Desert in Africa teach their children the importance of animal conversations, but in an unconventional way by Western standards. Jon had asked one of the San elders how this education begins, and he was told to return later that evening. He showed up to find the children of the village gathered in a circle before one of the older men. Prepared for a long recitation on predator-prey relationships, birdcalls, and other instruction, Jon was surprised by what he heard next. All the elder did was focus the children's attention and say, after a long pause, "A long time ago, the lions ate a lot of us... they ate a *lot* of us." Then he turned and walked away. The children's eyes widened and their heads swiveled in all directions. Those few simple words were all it took to prime them for a lifetime of keen observation.

In the developed world, we've insulated ourselves in a figurative rubber room with few sharp edges and no large carnivores. Jon maintains that bird language is one of the most direct ways for us to reconnect with our surroundings. True nature *connection*—which he is clear to distinguish from outdoor recreation and to separate from environmental education—comes only from direct involvement with nature's relationships.

Jon's distinction between connection, recreation, and education resonates with me. Whereas rock climbing, sailing, trail running, and studying nutrient cycles may have you out there thinking about the elements, unstructured time (preferably with an adult or mentor figure) in the outdoors is essential for children to build lasting personal relationships with nature. This is in large part why Richard Louv, in his widely acclaimed book *Last Child in the Woods*, highlights Jon's work as an example of what to do to keep children and adults connected to those elemental rhythms.

We humans are at our best when we get regular doses of those activities our minds and bodies evolved to do. We are infinitely healthier, happier, more creative, and more resourceful when we let the outside in. From Jon's experience, bird language and tracking are ways we can calibrate our entire being, from the physical and mental to the emotional and spiritual. When used to its potential, the body acts as an antenna for all that we experience. It is a permeable clearinghouse through which we absorb, process, and interact with this amazing world.

The following morning, we have breakfast, meet in the outdoor tent, and dig into the details of the class. Jon has a knack for bringing people along from knowing very little to a surprising level of awareness in the span of just one lecture, laying the foundation for what will be learned over the ensuing days. Jon first discusses the different calls used by birds. "There are two types of bird communication to be aware of," he says. "There is baseline and alarm." Baseline is what most of us would call normal, where everything is calm, relaxed, and easygoing. In baseline, the birds are singing, feeding, socializing, settling territorial scuffles, laundering their feathers, and caring for their young. Alarm, of course, is everything else.

Jon stresses that alarms can come in various forms; alarm is not a one-size-fits-all phenomenon. As he details in his book *What the Robin Knows*, there are a number of clearly discernible "shapes of alarm," as he refers to them. The shapes are postures or behavioral patterns that can be observed along with different vocalizations one might hear during times of unrest. There are clearly instances of over-the-top danger where everyone runs for the hills and seeks cover; this is what I first observed at the Brown Farm all those years ago. But there are other clues that are so subtle that most of us have missed them for years and would continue doing so unless someone pointed them out.

Jon states that "sentinel behavior" is among the most overlooked signals of danger in the bird world. What is sentinel behavior? Think of it like those armored chaps positioned atop a castle wall. They are the ones staring off into the distance, looking for the approach of imposing feudal lords and their bristling minions. Sentinel behavior is when a bird is sitting at the top of an elevated spot, not feeding or preening or doing anything else. It's sitting stock-still and looking in one direction. This could easily apply to any number of nonbird creatures too. Think of a neighborhood squirrel standing up on a rock for a better view, or even a fish pausing in place, pectoral fins waving.

Sentinel behavior also points to the importance of a sentinel's viewing posts. Would you ever guess that the dead oak tree in your yard that is slated for removal might be a key element in the local wildlife security system? There is a reason the herring gull or the bald eagle selects the highest boat mast in the harbor as a perch—it's not a small barstool idly taken up for a bit of rest, it's a lookout post. You might want to take a moment to check out potential lookout spots in your neighborhood or place of work. If you were on guard duty, where would you choose to hang out? Are any of these spots ever occupied? Now might be a good time to start looking.

As Jon points out, sentinel behavior usually delineates the edge of an area of alarm. We might replicate these same actions while attending a public event, when something unexpected goes down and everyone involuntarily stops to see what's happening. Birds, likewise, view situations from a safe position, and though this may not be as obvious as someone screaming bloody murder and running out of the room, it is still a state of alarm. Narrow down your search for lookout posts by thinking about where one kind of habitat abuts another—perhaps around the perimeter of a public park, along a hedgerow, or by a body of water. How is the landscape divided up in the places where you have chosen to deepen your knowledge of the wildlife around you?

With the help of a revolving body of students and individuals he has mentored, Jon has come up with several ways to distill nature's complexity down to some core principles, adding some of his own terminology along the way. The rush of birds fleeing the dog walker, or the wildlife exodus in advance of the Confederate infantry at the Battle of Chancellorsville, is something that Jon and his circle refer to as a "bird plow."

Kevin Reeve, a student of Jon's who teaches bird language to active-duty military, says that one of his students was on a rapid response team that was deployed after a convoy was hit by an IED (an improvised explosive device) in Afghanistan. The soldier used the bird plow concept to navigate this life-and-death situation. Using his tracking skills, the responding soldier quickly identified the bomb-setter's tracks. Starting near the wreckage of the deceased and injured servicemen and destroyed Humvee, he set out to follow the prints. As the enemy combatant had a substantial lead, Kevin's student employed a style of tracking known as "sign-cutting," a practice where the tracker leapfrogs ahead on a fresh trail to make up time and distance.

The soldier soon caught a glimpse of the enemy as he crested a hill and dropped into a small wadi or ravine. Moving quickly to that location, the soldier then noticed a bird plow off to his right, about three to four hundred yards away—the combatant, as viewed through the birds, had made a right-angle turn and was now going in a new direction. Knowing his enemy's new trajectory, with the aid of a map, the soldier reconfigured his position and hence the best point to intercept the enemy while keeping him at a safe distance. Using his bird-sourced intel, the soldier was able to complete the mission for which soldiers are ultimately trained.

Jon and his colleagues have coined several other terms to define specific types of alarm behaviors, including "popcorn," "hook," "ditch," "bullet," and "parabolic." Each of these labels describes a physical response of birds to a given disturbance. Generally speaking, each type of threat elicits its own specific brand of acknowledgment.

If you see a fast-moving predator like a falcon scaring birds, you will see the birds exhibit a speedy flight response in a unified direction referred to as a "bullet." When there is a snake in a tree or an owl perched deep inside some evergreen foliage, the birds dance around the static focal point in a formation that resembles a feathered stress ball. This is known as an "owl" or "nest robber" alarm. For a larger static danger on the ground, birds will hop up to a safe height above it and focus their scolding downward, positioning themselves in an umbrella formation known as a "parabolic" alarm. The term "popcorn" describes the way birds leap up out of the undergrowth and seek safety in the air or on branches in response to animals moving about on the ground.

Try observing how birds respond if your cat manages to escape into the outdoors. They will likely flutter skyward to escape Felix's claws and teeth, but only as high as they need to and no higher, as excess movement would be a waste of precious energy. A bird's height off the ground can help you figure out who's in the mix. "With practice," Jon tells us, "you can know with a high degree of certainty what the threat is and, in some cases, even the exact individual involved."

Jon explains that when you add vocal tidbits to these shapes of alarm, you begin to get a clearer picture of what is going on. Taking stock of certain calls, phrases, or tonal inflections can be a real asset as you drill down on the details. When I first learned about these little gems from Jon's tapes, they helped me clarify observations I had made while growing up but had never fully understood, much less articulated.

After Jon's late-morning lecture, we eat lunch and then break into groups. Each group is assigned a leader and a discrete section of the event grounds to use throughout the program. We soon head out for a walk to familiarize ourselves with the surroundings.

The landscape is a patchwork of gardens and mown and unmown fields bordering a white farmhouse and associated out-buildings. The open ground is rimmed with forests of redwoods and Monterey pine. It could be almost any pastoral setting were it not for the distant sound of the crashing waves of the Pacific Ocean to the west. A creek flows in a vegetated gully just beyond the house, and a light breeze moves from southwest to north-east—it's a beautiful place to spend a few days.

My group leader's name is Guy. Guy is a few years older than me and has piercing, empathetic blue eyes and a broad smile. He has been mentored by Jon for several years. Our group gathers to learn the logistics of a "bird sit," which will serve as the prin-cipal hands-on instructional tool for the workshop. I mentioned ways to approach a "sit" in an earlier chapter. The key difference this time is that the sit is done with several other folks and the time is divided into five discrete intervals over the next hour or so. Because we are not allowed to use our watches or phones—in order to exist in a state of relaxed timelessness—each interval, 0–4, will be demarcated by a crow call given by one of the group leaders. This will allow us to sync our sightings afterward.

With foam sitting pads, binoculars, and notepads in hand, our group spreads out along the edge of a cut field east of the farmhouse. Guy instructs us to select a cozy spot, the only requirement being that we remain still, quiet, and watchful. For my post, I select a seat on the ground along the field edge.

Ankle-high grass stubble surrounds me, and taller grass, shrubs, and young trees stand to my back. Laid out before me is the open acreage of the workshop grounds, including our sleeping tents, the kitchen area, the lecture tent, and the parking lot. It is roughly 2 PM and hardly an ideal time to watch wildlife, as the heat of the day often drives many animals into cooler cover. Although group members have ended up no more than fifteen yards apart, I realize the experience of this place is entirely my own. Some birds are active, and I jot down my observations under the appropriate time slots bookended by the imitation crow calls. But more than anything, I breathe deeply and relax in the warmth of the sun.

A sparrow moves along a grassy hedgerow down by one of the greenhouses. A crow flies overhead. There is an olive-sided flycatcher singing its *quick-three-beers* song near the creek. I can hear other birds behind me and to the right; some of them I know, but many of them I don't. This is my first time being around many West Coast species, including golden-crowned sparrows, Bewick's wrens, and bushtits, and though I studied recordings beforehand, the sounds didn't quite stick.

Before I know it, I hear a double crow call signaling the end of the sit and we reconvene as a group back down near the lecture tent. Guy has us circle up in the sunshine as he spreads out a large sheet of paper and drops a fistful of colored markers next to it. In black, Guy scribbles in landmarks to make a rudimentary map. Then he hands the black marker around for us to help fill in some of the gaps: the driveway here, the house there, the creek on that side, a hedgerow leading to the south on the other, this tree here, that isolated shrub there.

"So, what did you see and hear?" Guy asks.

We discuss each time period one by one, starting with time zero. This is when we walked into the sit spot area and were most disruptive. The discussion starts off slowly, but as the group gets comfortable, the ideas begin to flow. Many noticed birds fleeing the open meadows and disappearing into the brush during time zero. These bird plows were small to be sure, but they were bird plows nonetheless—followed by a universal hush. You may have noticed a similar pattern when doing your own version of a sit; the activity often starts to pick up toward the later time windows.

The utility of sitting so close to one another becomes evident. It is astonishing how different everyone's observations are despite our proximity. I always felt like I had a good handle on what was happening around me, but the group sit showed me how much I was missing. This one activity cemented my understanding of why studying animal conversations with a group is so beneficial. Things I saw and heard were missed by others and vice versa. Each one of us was looking at slightly different facets of this same big jewel.

One by one, the group members flip through their notes and scribble them down on an increasingly colorful, cluttered map. Things that happened in time zero are noted in blue, orange signifies time period one, red for period two, green for three, and purple for time period four. Although the picture on the paper quickly becomes a mess—with sparrows doing this over here in orange, robins saying that in red over there, and a jay squawking in green somewhere else—our collective understanding of what was going on during the last hour begins to take shape.

"Did you see the sparrow?" asks a voice in the circle.

"No, but I did see the flycatcher," another replies. "I think it was an olive-sided flycatcher; it buzzed me, grabbed a bug, and then landed over in the shrubs on the edge of the trees."

"Did you guys hear that dog barking in the distance?" another voice inquires.

"How many jets flew over?" Guy asks.

"Jets?" says one of the women. I could almost hear her thoughts. "Why on earth are we listening for jets in a bird language class?"

I knew where this line of inquiry was going, so I played along. I offered that I had heard three. Someone else chimed in that they had heard only one, but that they saw it too.

"Four. I heard four," says another with a degree of confidence.

"There were seven," says Guy as he looks knowingly at the group.

"*Seven?*"

"Yes," Guy replies. "Seven."

The reason airplanes are brought up is that in order to really decipher the language of our feathered friends, we have to be open to everything. Why? Birds can't pick and choose their soundscape. No animal can ignore unpleasant aspects of area noise for the sake of convenience the way we humans are accustomed to doing. Everything overlaps and intermingles. The tiniest clang or peep can radically change an animal's behavior, and the noise and the animals' reactions all influence what we see.

In his book *The Great Animal Orchestra*, Bernie Krause, a California-based audio ecologist, described the unfortunate fate of some Great Basin spadefoot toads at Mono Lake. The jarring sound of a low overflight of a military jet silenced the toads'

nighttime chorus for thirty or forty-five minutes. While the pro-
tective wall of sound provided by the full chorus was in shambles,
a pair of coyotes and a great horned owl slipped in and picked
off a couple of the toads that were trying to reboot the ensemble.

The impact of human-generated noise on wild animals
should never be underestimated. Later in the same chapter,
Krause shares an account of a jet flying low over a zoo north of
Stockholm, Sweden. The aircraft so upset the lynx, foxes, and
tigers that they killed and/or ate twenty-three of their own
young, including five rare Siberian tiger cubs. The next time you
venture out to create a sound map in your neighborhood, be sure
to include human-made noises as well as natural ones, as they
may well be the key to figuring out what is going on—though
hopefully with less dramatic consequences than the examples I
have just given.

Guy then poses a question that flips my lid. "How did you feel
during the sit?" he asks.

Right away, my New Yorker sensibilities start flashing. I chalk
the question up to some new-agey Californian thing. "Here we
go," I think. "Are we headed for trust falls and sappy confession-
als?" But Guy is after something different, and I am converted in
a matter of minutes.

Feelings are the tip of our emotional icebergs; they are the
clearinghouse for our subconscious, which is constantly detecting all that is within and around us. What the group has been
pouring into the conversational mix from our notebooks, as we
wield colored markers, are constructs of our conscious, thinking
minds. What Guy is after now is a deeper assessment of the scene,
an assessment that channels into our psyches through hidden
pathways.

As we reflect, even the newbies to this whole bird-language thing come to many of the same conclusions as the more experienced birders, despite not being able to tell a vireo from a cockatoo. Top-tier identification skills, as it turns out, are not necessary when it comes to getting the gist. It is possible to sidestep identification and classification in favor of yielding to *sensation*. It's not unlike how we feel our way through a Bach concerto or an emotional conversation; there is something deeply ingrained in us that passively absorbs impressions from the environment. Jon will later refer to this interaction between the subconscious mind and our intuitive selves as a kind of "extended cognition."

"Yes. Back behind us to the right," someone says. "There was something going over there."

"I felt it too," says someone else. "There was definitely something going on in that corner of the forest."

During the sit, I had felt calm for the most part, relaxed, but there was a point when I had found myself moving around in my seat. Craning my neck, I found myself scanning for something I could not entirely describe or pinpoint. It came and went. I gave it little thought, but something had nudged my inner barometer.

As responses to Guy's question filter in from the others in the group, they too agree that there was "vaguely something" going on during the third period. And indeed, there it was for all to see, as outlined in green marker. Whatever was going on to the northeast of us, it left its impression on all those present. We felt it.

Guy nudges a bit further. "If you had to write a newspaper headline about what went on during the sit, how would it read?" Someone in the group offers up, "Weird stuff in the woods over there." Heads nod in affirmation.

A short time later, we mingle with members of another group who happened to be sitting over in the area in question. One of them shares that she had seen what looked like a small hawk dart through the trees at the corner of the field and disappear. The group had seen smaller birds fleeing in all directions and sounding out alarms that lasted for several minutes. Another student from that group had spotted a sharp-shinned hawk; it swooped down into the mid-level trees, nearly catching a bird, then looped back around and flew to the east not fifty yards behind our group.

We complete several more bird sits over the course of the five days, but, more importantly, we do class-wide debriefs with all the other groups. Input from the entire student body is compiled in these sessions. The group-wide effort exposes a more richly detailed picture than I could have ever imagined. Each time we sift through the particulars of a given sit, I can see awareness of and proficiency with bird language building in all of us.

If a picture is worth a thousand words, each one of us has an hour-long multimedia extravaganza during each of these sits. After combining our narratives, it is clear that one person adding their bit to the collective is not additive; rather, it is synergistic. One plus one does not equal two; in this case, it equals a thousand. Not only does a bigger image emerge, but it becomes a living, breathing story. And with everyone's eyes, ears, and minds focused on a unified goal, seventy independent stories merge into one.

Each time we debrief, we puzzle out fascinating chains of events that occurred in exactly one hour. What else is going on for the remaining twenty-three hours of the day, I wonder? Then a thought occurs to me: What if we were part of a community that sought out and valued these sorts of stories hundreds, if not thousands, of times each year? And what if that composite narrative reached back not just one year, but millennia? What are we missing? I become dizzy with the possibilities. This is part of the intangible brain trust that disappears when one culture deposes another.

Jon later shares another story that addresses this point. On a sweltering afternoon in Africa's Kalahari Desert, when he and a traveling companion were visiting the small group of San people I mentioned earlier, Jon's friend noticed an outburst of bird alarms near their wooded encampment. Being keen students of animal language, Jon and his friend identified this noise as a signal of danger—likely a snake—about 170 yards away. Having been warned not to stray from the forested patch and into the open savanna unaccompanied, they did not go to investigate. Instead, they called upon one of their translators, an astute student of the bush in his own right. They asked him if he thought the sound might indicate a snake in the distant tree. After walking about fifty yards closer to hear the sounds better and to have a look, he said, "Yes, I think it is a snake—maybe a mamba, or perhaps cobra, but let's go get Issiqua." Issiqua, one of the San men whose name literally translates to "cobra" in English, would know better. Minutes later, and in Issiqua's company, they walked a few yards closer than before, and from a distance of about thirty yards he confirmed, "Yes, it's a black mamba, don't go over there."

When they returned to the forested encampment, five women were seated together in a half-circle, making beads out of ostrich eggshells. After waiting patiently until the women were done with their conversation, Jon asked through their translator, "Are you hearing any alarms?" In unison, all five women pointed directly over their shoulders behind them, toward the snake-bearing tree, and said, "There's a mamba, don't go near it." While conversing with each other, making beads with their backs to the serpentine drama, surrounded by dense trees, nested within the clamor of men talking and laughing, tanning hides, and in the company of playing children, these women had been listening all along. Not only had they heard those faint background bird alarms from almost two football fields away, but they had interpreted their *exact* meaning.

In San culture, the women are among the most astute practitioners of animal language because they are commonly gathering food and materials in dense vegetation, in places with dangerous animals, and in the company of children and elderly people. The San are the descendants of the first modern people in the cradle of humanity, and this was but a small indication of the cultural inheritance derived from living in a place for a hundred thousand years.

My observations during that bird conversation workshop in California are not as dramatic, but during the final bird sit, I see the key players around which the rest of the sit's conversations take shape. The whole drama unfolds right before my eyes. Although I miss the initial movements, Jon catches them while he's down in camp. He sees a mid-sized bird drop to the ground as a peregrine falcon rockets down out of the sky, nearly striking the bird before ditching into the grassy turf.

I spot the falcon just as it rises from its stoop and heads north after the near miss. The falcon causes the forest to erupt in panic. Alarm calls ricochet from all quarters, and birds duck for cover wherever they can. Some freeze in place, hoping that by becoming birdsicles they will not fall prey to this most lethal of hunters.

I watch as the stiff, sharp wings of the falcon turn toward the ocean, then sit transfixed as the raptor flaps farther and farther away before becoming a tiny speck on the horizon.

At the class-wide debrief, there is a palpable energy among the students and banter is lively and fast-paced. People talk in excited tones and elevate their voices; they wave with their hands and involuntarily mimic the ways in which some of the birds moved. I confide to my group what Jon and I saw, but we hold our cards until the debrief reaches its apex. A headline might have read: "Something Big Happened—Guess Who." How did we feel? The sit had felt tense, jaw-clenching. Even those who didn't see the falcon intuited its presence. Is this where notions of a sixth sense are rooted? We felt the series of events not as dispassionate, objective observers, but as if we were one of the feathered masses—we were rewilding ourselves.

In the end, I finally have the chance to shake Jon's hand and thank him for all he contributed to my understanding of bird language all those years before. Harnessing the power of multitudes is the fastest way to advance your skill for eavesdropping on wild conversations. Round up a group of friends—even if they know nothing about birds and you have to ply them with drinks and pastries—and organize your own sit. Involving others opens up a timeless narrative that is greater than the sum of its parts.

- 17 -

THE HONEYCOMB LANDSCAPE

F YOU DON'T MANAGE to pull together a gang for a group bird sit, make sure you get out anyway. There is always something to learn. Undecipherable messages will continue to hover on the periphery of your consciousness, but keep going. The more you experience, the more you will feel, and the more you feel, the more you will connect. What additional information can you gather? And what if you did this for a long time—like, a *really* long time?

Author and biologist David George Haskell visited one patch of old-growth forest near his home in Tennessee every day for a full year and chronicled a rich trove of discoveries in his book *The Forest Unseen*. To learn much of what he knows about bird language, Jon Young sat at the base of the same ash tree each day for seven years. Rick Bedsworth has done something akin to Jon Young, but in his own unique way. I share Rick's story with no expectation that you will drop everything and follow suit. Rather, it is to show what is possible when curiosity and persistence are allowed to reign supreme. As we will see, nature is better than any video game; there are countless levels of skill to aspire to, and the game is never over.

Rick Bedsworth was introduced to me by a mutual friend. He is a focused and deeply inquisitive man with cropped brown hair and a stout, fit build. A talented athlete during his high school years, Rick was a standout baseball player. He had little difficulty walking into tryouts for pro baseball teams right after high school, but as a free agent, his contract options offered only a pittance to support his young family, so he joined the workforce instead. After a position he had taken in Washington state fell through, Rick moved back home to the Kansas City, Missouri, area with his wife and young daughter.

Rick had been a kid who played outdoors with his two siblings, and he is quick to tell you that he was no prodigy when it came to school or other studies. What he did possess, however, was a powerful drive and the ability to focus on routines that allowed him to succeed in areas that interested him. Every day, before heading to work at his father's trucking company, Rick would drive to his special sit spot and get into place well before sunup.

He continued this routine day in and day out for almost twenty years. He would go on weekends, in the evening, in the middle of the day, through rain, wind, and occasional snow, sometimes making two or three trips in a single day. Rick wanted to see this familiar place in all its iterations. He wanted to know its true character. At times when he was on the road for work or he couldn't be at his regular sit spot, he would pick out a nearby location that could give him some of the same exposure. And what he learned at home, he found, applied in great measure to these other spots along his travels.

What do you discover after sitting under the same tree through all manner of weather and seasons for two decades? The answer to that leaves me giddy; he learned things that few, if any, postindustrial humans would believe. Going solo meant things did not always come as quickly for Rick. It took him three years to realize that the persistent lack of birdcalls around his sit spot was due to a nearby Cooper's hawk nest.

During his daily vigils, Rick committed to sitting for at least thirty minutes, but preferably for an hour or more. Although most of us probably cannot muster the time for this level of commitment, it is instructive to know what can happen when you do. Rick is fully transparent and discloses that many of his greatest epiphanies came only after many years of field time, and in some cases after over a decade of continuous viewing.

As we talk over Zoom and the conversation shifts away from his work and upbringing and into his wildlife encounters, Rick becomes visibly animated. Many of the things that he uncovered were not things he was looking for per se; they sort of just occurred to him. Reticent to draw hasty conclusions, Rick would discover numerous patterns, but continually hedged against snap interpretations. He framed them as "Well, this is what I've been seeing and this might be one explanation. But I'm going to keep watching." And so it would go. Rick continued collecting evidence and more data to test his hypotheses. Only after many iterations, follow-ups, and still more time at his sit spot would Rick concede to making up his mind.

In the beginning, everything just looked green and somewhat two-dimensional. Over time, Rick recognized how forest diversity, the presence of edges and transition zones, was key to understanding what he was seeing. From his post beneath the layered canopy on a prominent ridge, he gained an appreciation for the importance of differing tree age, slope, aspect, and weather, and how all the living and nonliving things interacted against the backdrop of seasonal phenology.

Something as simple as a leaning tree—pulled over by gravity and the inexorable weight of wild grapevines—becomes a valued larder for all sorts of creatures in autumn. This sort of tree could easily be a target for the woodcutter's saw, but it serves a decidedly different function from its upright neighbors. The connections between the land and its inhabitants are myriad and deeply fascinating to Rick.

One thing that stymied Rick was the density of the vegetation. It was hard to know what animals were telling him when trees and shrubs blocked his view. Who might be responsible for events in the outer reaches of his awareness? Many of those answers came only after he became a better tracker. Taking workshops with teachers who performed search-and-rescue operations or tracked illegal immigrants along the Mexico–United States border helped a lot. Rick learned how to follow animals across challenging surfaces such as hardpan earth and leafy forest floors. Instead of finding clearly delineated prints in sand or snow, he had to reorient his search image for softer, less discrete impressions in woodland understories.

Improved tracking skills soon helped Rick follow local deer with ease. He might realize there was not one, but maybe two or five individuals using an area, and that this was a family group distinct from another assemblage a short distance away. As Rick followed the deer, there were times he might inadvertently nudge them from their bedding locations. Through the bird language, he could decipher that they had snuck away in a thicket off to one side—denoted by the alarm chip of a cardinal—and a broader picture began to coalesce in Rick's mind.

He became adept at reading alarm and baseline states, but he pushed those interpretations to new heights. From his sit spot, he tracked bird plows, sentinels, bullets, parabolics, nest robbers, and other shapes of alarm. Rick came to know the signatures of each animal species the way a sommelier might disentangle the mineral notes, tannins, and pH found in a glass of wine. Any number of cues allowed him to predict that a Cooper's hawk, not a crow, was about to show up many seconds before it actually did. Reading the ripples in the living waves around him, he could identify the approach of a raccoon well before the masked creature entered the picture.

In particular, Rick came to recognize that silence was the most common alarm of all. Silence, as it turned out, is much more than the absence of sound. Silence is brimming with meaning. The forest inhabitants notice it too. The outdoor hush is not uniformly distributed either; it exists in pockets. Using his refined tracking skills, Rick could verify that one pocket of silence surrounded a perched owl or a weasel's den. Another muffled sphere might bound the nesting site of a hawk, and coyotes and bobcats added their own signature brands of quiet. As a threat moved, the silence followed.

At a certain point, it became clear to Rick that the animals in the forest recognized the seriousness of silence. Potential prey species navigated around the acoustic dead zones to avoid trouble and lingered in places where the neighborhood watch was still intact. Outside the black holes of silence, the sparrows were singing, the cardinals were feeding, the woodpeckers were tap-tap-tapping away, and the dark-eyed juncos were doing their two-footed kick-scratches in the leaf litter as they foraged.

The predators, in contrast, were piggybacking off one another's interference patterns. Rick found that when a bobcat encountered an area of silence, it would veer directly into it. With the neighborhood watch shut down and the gossips no longer vocalizing, the bobcat could make its way through the forest unharassed by birds and small mammals. This radio silence allowed the hunter to fly under the radar when it would otherwise encounter a riot of alarms on untrodden ground. When the cat exited the first patch of silence, it would then direct its movements straight toward the next-closest area of silence, again masking its presence. After about twelve years of continuous watching and seeing these patterns repeat over and over, Rick codified his thinking. To him, this looked like one enormous landscape-scale game of hopscotch hide-and-seek. He called it "the honeycomb landscape."

The events of one particular day catalyzed Rick's thoughts. Shortly after a Cooper's hawk flew into the area and triggered alarms and the inevitable ensuing hush, Rick heard a distant rustling in the leaves. Whatever this was, it was coming closer. It turned out to be not one but three deer—a mother and two fawns. As the trio approached the area of silence triggered by the hawk, the fawns continued on their trajectory and waltzed straight into the sonic void. Rick turned his attention to the mother, who was bringing up the rear; he was shocked by what he saw next. Upon reaching the edge of the pocket of silence, the mother deer stopped so abruptly that a visible wave of flesh rippled through her body from front to back. It looked as though she had smacked directly into a plate-glass window. The mother came to a stop as definitively as a pedestrian might when noticing a speeding car as they were stepping off a curb.

The switch in the doe's behavior when leaving the safe "wax" of the comb and butting up to the edge of the silence was a sight to behold. Poised on rigid legs, the mother bleated once to get the attention of her heedless youngsters—they stopped. She vocalized a second time to call them back to her side, and they dutifully returned. Once reunited, the three deer departed in the opposite direction. It was at this moment that it all came into focus, and Rick realized the true importance of nothing—silence was indeed something.

The doe's reaction to a plot of quiet forest ground was a revelation. We think of animals walking on predictable routes. A trail goes from this area to another. They may diverge to navigate a river crossing, sidestep a highway, flank a housing development, or follow the edges of farm fields. Yet here we have the honeycomb landscape, a completely invisible influence on animal movements, one that directs them as strongly as any physical feature out there.

From that point forward, Rick would continue watching at his sit spot through the lens of the honeycomb landscape. What he found reinforced his thoughts. Rick had unearthed an ephemeral, moment-to-moment messaging machine that was continually shuffling the deck. Instead of going from A to B in a straight line, or even along a trail, animals of all sorts could be seen diverting away from paths and recognizable corridors to follow the edge of silence. As Rick's awareness of these and other phenomena grew, he could monitor an area of perhaps two-and-a-half miles across and know what was happening, in real time, almost anywhere within that circle.

Interestingly, when other people joined Rick, they did not gather the same impressions. Even when they were walking right next to Rick, other observers did not detect the shifts and patterns he was noticing. Although they could see and hear the same signals he did, they did not recognize the meaning behind those messages.

Domestic animals, like their owners, seem entirely oblivious to the honeycomb landscape. Over the span of several weeks one autumn, a lost yellow Labrador was inhabiting the area around Rick's sit spot. Instead of behaving like a wild predator and seeking out the nearest silence, the dog blundered straight through the honeycomb matrix from one zone to the next. Before long, the pooch was worn out and hungry. The dog behaved like a manic off-road vehicle with no regard for accepted etiquette or travel routes. Rick could easily identify the location of the Labrador, and in a few cases, he ended up scaring the bejesus out of the poor dog by showing up where a human wasn't supposed to be. Rick was able to briefly befriend the dog by bringing it much-needed food until it was finally reunited with its owner.

When I think of the honeycomb landscape, I recall an experience I had with some schoolchildren in northern Yellowstone many years ago. We were at the Buffalo Ranch in Lamar Valley at lunchtime when a boy came sprinting in from the back porch, yelling at the top of his lungs,

"Come here! You gotta come see this! You gotta come here *right now*! Hurry! *Hurry!*"

Everyone stopped what they were doing and quickly followed him to the back porch. As we spilled out the doorway, he exclaimed again, this time pointing to something on the open valley floor.

"Look, look! Do you see it? *Do you see it?*"

Everyone was on pins and needles, and yes, we did see it. It was huge! An enormous dark mass was surging across the valley with incredible speed. It was larger than any buffalo herd I had ever seen, and it swept overland while consuming everything in its path. It engulfed sagebrush and the meanders of the Lamar River and then Rose Creek, and soon overtook the road and kept spreading.

By this point the boy was spellbound. He was gripping the hair above each ear in his fists in a state of shock. Several other children were frozen too. They were incapable of doing anything but staring—it was their first cloud shadow. This group of urban children had never been in an open, undeveloped landscape that allowed them to see the shadow of a cloud fully express itself on a horizontal plain.

I think of that day every time I see a well-defined cloud shadow, but I also think of the children's reaction as a metaphor for the honeycomb landscape. When we think of landscapes as static and existing only within familiar parameters, we limit our perceptions of what is and is not possible, and as a result are sometimes blindsided by what reality offers. The whole cloud-shadow episode was as fantastic to those children as the most masterful magician's trick, yet it was real. It was as real as it gets.

In time, Rick was able to use the honeycomb landscape to his advantage. When walking to his sit spot, he would adhere to his newfound rules of the forest. By walking inside the waxy safe zones of the comb, he could move freely as other forest dwellers did.

When I ask Rick how people might be able to identify the invisible line between normal and the eerily quiet holes at home, he suggests watching the sentinels. He tells me they are the most informative signposts at the edges of silence. A stationary gull on a ship mast or a chipmunk perched on a log is, itself, safe—but it is almost always looking into an area of instability. In other words, a sentinel often rests at the edge of the danger zone while keeping its back to the calm of the baseline. With their two toes forward and two toes back, woodpeckers perch on the sides of trees like the handset of an old wall phone and act as perfect sentinels. In their elevated positions, Rick tells me, woodpeckers are one of the best indicators of the edge of silence in his area; in essence, they act as the crossing guards in the honeycomb landscape.

Rick's newfound knowledge helped him blend in with the locals. Chipmunks, wrens, and robins saw in Rick's behavior a sensitivity to silence that mirrored their own. Rick could literally walk through the forest as a cardinal was singing directly over his head. Goldfinches fed in loose groups all around him, and in one case, a raccoon arrived at his boot tips, looked up nonchalantly, then proceeded to its den tree in full view of Rick's wide eyes.

If Rick did disturb a towhee or a junco feeding on the ground, he would stop. Rather than pushing them to the point of alarming, he would avert his gaze and wait. Once the animals settled down, he might even speak in a soft voice, saying something like "Sorry, I didn't mean to bump into you there." Only then would he proceed. Animals respond positively to this sort of deference or, as some might call it, politeness. Rick's walks evolved from silent forays to lively outings filled with all manner of wildlife activity that carried on all around him as if he were nothing more than another fixture in the woods.

Early peoples lived in the open environment their entire lives. The collective brain trust of ancestral knowing, compounded by a hunter-gatherer's own need to survive, emphasized the need to pay attention to minute details. Rick's experiences seem to echo those time-honored traditions. But how much deeper do phenomena like the honeycomb landscape go? Unfortunately, the answer is unavailable to us, but it is a beautiful thing to think that if we are willing to put in the time, rediscovering transformative relationships is most certainly possible. I leave the conversation with Rick energized in a way I have not been in a long time—and with a bevy of things to start keeping my eyes and ears on the lookout for.

"There are no shortcuts," as Rick would say, but there are some bridges that can span gaps in our knowledge and abilities. There is absolutely no substitute for putting yourself out there. If you want this kind of harmonic rapport with nature enough, if you want to interact with the wider world on its terms, pick a good spot to sit and visit there often. Do your best to identify a patch of silence if you can, even if it is the one you created when walking to your spot. Can you delineate the edges of an area of silence? How big is it? How long does it last? Does it move? Even if you don't ascend to the level of Rick Bedsworth, there is a lot to be gained as you seek the sweet interior of the honeycomb landscape.

- 18 -

WILDSENSE 2.0

T HERE ARE COUNTLESS WAYS in which we frighten wildlife, but we can begin to shift that dynamic by building trust. One way to start out on this path is to extend common courtesies to nonhumans. Don't blunder heedlessly into their quiet, restful scene. Avoid direct eye contact and let your body language do the talking. Move slowly with a heightened sense for everything around you. You will likely be surprised with the rapport that even small gestures can build—it can be a game changer.

Let's say you are walking and encounter deer along the trail. Rather than continuing on your route, stop. Pausing is the first step in diffusing the tension triggered by your arrival. The deer still might melt into the background, but if they don't, turn at an angle away from them and keep watching them using your less confrontational peripheral vision. It's a bit like the childhood game where you cover your eyes and pretend that if you can't see the other person, they can't see you. When you do this, many animals will feel as though they are still hidden or, at the very least, feel more comfortable if they know they've been spotted. Jenny and I do this all the time and are often able to walk past deer and elk with our dog, and they will continue grazing.

When you startle some deer and then stop, you begin a nonverbal dialogue with that animal. You are saying, like Rick Bedsworth, "Sorry for bumping into you. Please relax and continue on as before." Remain still. Maybe even take a step or two backward. Watch how the deer respond. Have they calmed further or are they still on edge? Minutiae become critical at this stage. Look for delicate shifts of weight from one foot to another. How animals carry their heads can say a lot. A deer that holds its neck straight up or stretches it out horizontally to its full length is likely tense. The position and movement of its tail are another important bellwether.

Assuming the deer have not run away, you now have a chance to enter the next stage of the conversation. If the deer were traveling when you first encountered them, give them time and space to continue. If the deer are not on the move, wait until they begin feeding or grooming before you start to move again. Other cues that a doe or buck is calming down might be that their ears and tail are not quite so erect; they may begin chewing their food again. Observe as their eyes, legs, and overall appearance begin to relax. Each species has its own gestural cues. Look for these species-specific markers of relaxation and commit them to memory. You can learn some of these behaviors from books and classes, but time spent together will be your best teacher. The more time you spend practicing these techniques, the more in sync with nature you will become, and the more trust you will build. Rick Bedsworth has shown us, with more trust comes greater leeway when moving about.

As Jon Young explains in *What the Robin Knows*, the goal is to expand your sphere of awareness while contracting your sphere of disturbance. When you move along the trail, acknowledge wildlife's personal space and give them extra room. If animals appear uneasy, stop and repeat the same sequence you used at the start. Jon refers to this practice as the "honoring routine." In the initial stages, you may try approaching the situation from the three-strikes-and-you're-out principle.

What is the three-strikes principle? Kevin Reeve, who instructed the soldier who served in Afghanistan, brought the importance of watching animals to the fore during a training exercise for a group of marines at Camp Lejeune in Jacksonville, North Carolina. Given that the military is built on a foundation of strict protocols and technologically advanced equipment, birdwatching skills are not often on their radar. Kevin changed their minds in a matter of minutes. He did not employ words to illustrate his point; he showed them.

Kevin began by leaving a trail of his own footprints while walking into a patch of forest. He then turned ninety degrees within the trees and walked parallel to an adjoining field. Halfway along the open meadow, he left the trees and returned to a parked Humvee. The marines, who were shielded from seeing Kevin's walking route, were instructed to track where he had gone. While the marines followed his trail, Kevin visited at the Humvee with the NCO (noncommissioned officer) in charge. At one point, Kevin suddenly stated, "They're about to come out of the field in a minute or two, right near that one tree."

The marines had lost Kevin's tracks and were aborting the exercise. As if on cue, the marines stepped out of the forest right beside the tree Kevin had identified. The NCO turned to Kevin, flabbergasted. How could he have possibly known where the men would exit and when? Had the NCO already been a student of animal language, he would have known. A bird plow was leading the way. "All you had to do was look for the birds," Reeve said.

Once the marines returned, Kevin introduced them to the three-strikes-and-you're-out principle as it applies to birds. Kevin walked the men back into the forest and stopped at the first bird they encountered—a sparrow. "See that bird?" Kevin said. "If you scare him, you scare everybody, but you get three chances." Instructing the men to stay put, Kevin advanced toward the bird. The sparrow switched its position on its twig ever so slightly and raised its head. "That was strike one. Now watch." Kevin took another step forward and the bird jumped to a higher twig, flicked its tail, and made a few furtive calls. "Strike two." When Kevin took one more step, the sparrow flushed and gave a loud, chipping call. Strike three triggered the expected cascade of alarms and opened up a silent hole in the forest-animal conversations.

Where the NCO and the marines had previously been surprised by Kevin's superhuman abilities, they could now see that the pattern was repeatable, reliable, and incredibly useful. Three-strikes-and-you're-out is the equivalent of the ABCs and 123s of the woods.

The more animal signals you recognize, the more respectful you can be toward the animals you encounter. Deferring to your local pigeons or crows around town may seem a bit odd to some, but to be accepted in wider circles, this kind of conduct is essential. Sometimes you may have to walk through briars and thickets, bogs, or deep snow to get around those you run into, and I'm sure you might be thinking, "Oh my gosh, I'm never going to get anywhere at this rate!" And that's right, you won't. The sooner we stop pushing animals' buttons and start allaying their fears, the more quickly we can establish a relationship of give and take. The destination is not a physical place in this case; it's a state of being.

To be fair to yourself and others, it is best to practice the respectful walk when there is no agenda. Pick a weekend when you are free to roam for as long as you like. The need to be somewhere at a certain time is what shifts the human animal into predator mode—at least, as far as potential prey are concerned. There are times for covering ground, getting exercise, and climbing peaks, but there are also times for wandering.

Jon Young tells a wonderful story of being out on a walk with a group and, on the way back to the parking area, encountering a flock of Canada geese. Instead of walking right through the geese and flushing them, he applied the principles of the honoring routine. By taking a few steps forward, then pausing and signaling the group's intention to return to the parking lot, he allowed the geese to adjust on their own time and on their own terms. The result was that the birds remained calm. In what seemed like a touching gesture of understanding, the flock parted into two, opening up a passageway leading back to the parking lot. Jon instructed the participants not to look directly at any of the geese or stop or take any pictures. These little acts would be a breach of the birds' trust. The group agreed to the terms and walked straight through the flock without ruffling a feather.

Practice the honoring routine enough and thrilling things may start to happen. It took Jon Young several years of honoring the animals around him before he had the experience of walking to within a few feet of a fox on a nighttime outing. We might say that Rick Bedsworth, Jon Young, and Joe Hutto have achieved WildSense 3.0, 4.0, or higher, but an odd phenomenon has come to the attention of these individuals. If you have an "Oh my gosh, nobody's going to believe this" thought, it's over. It is not clear what switch is thrown, whether it is olfactory, visual, or even just mental—an experience I have had and will illustrate in the next chapter. Whatever the means of transmission, animals detect the change.

Respecting animals also means being in the moment. The French philosopher René Descartes famously wrote, "I think, therefore I am." In this context, a crucial word is missing. It should read "I think, therefore I am *distracted*." If you are out of touch with the present and caught up in your thoughts, where are you? Many have had the experience of talking to a friend or loved one only to realize that the person to whom they are directing their conversation is not actually listening. The cognitive disconnect is as troubling to wildlife as it is to us.

Being distracted means that you miss things. Consider the famous invisible-gorilla study, where participants were instructed to watch a video and count the number of times the players in white shirts passed the basketball. Midway through the video, a person in a gorilla suit walks through the game (you can see a video of it on YouTube), stands in the middle, pounds their chest, and then exits. "Did you notice the gorilla?" study subjects were asked. More than half of the observers missed the gorilla entirely. A similar lack of attention happens when we are focused on identifying a bird, caught up in our internal dialogue, checking text messages, or trying to get somewhere in a hurry. Slow down to the pace of real life—slow down to nature's pace.

There is no way to fully quantify an attentive state, and no way to bottle it up and pass it to someone else; we know it when we *feel* it. Each one of us is an individual, and our perceptions of the world around us are entirely unique. However, we share one thing in common: persistent exposure helps us slow down and put ourselves on the same wavelengths as our wild friends. Acceptance comes from practices of heightened awareness and a lack of fixated agendas. If you practice being in nature with a respectful demeanor, you will find you are more welcome among your wild neighbors.

Repetition of being in the moment and opening up to respectful practices will shift your neural networks toward thinking and acting in more considerate and compassionate ways. How long will you need to practice the honoring routine before you can mingle more freely among the forest's or estuary's residents? It will depend on the time you have available and your degree of dedication. Keep in mind that there are certain places where practicing this sort of activity will not be permitted. National parks, as one example, often stipulate buffer distances between humans and wildlife. Know the rules. Where regulations prevent close company with the more-than-human, private lands will be more suitable.

Regardless of the rules, make a practice of being sensitive to the animals' moods, no matter the distance you are required to keep. If you are not sure how things are going between you and your wild neighbors, check back with your senses. Let the animals guide you; they are continually projecting updates on how they are feeling. They telegraph their discomfort or sense of ease to us in very concrete ways, but only if we are willing to slow down and be more sensitive to what they are saying.

- 19 -

BECOMING
INVISIBLE

RACTICING REGULAR ACTS of politeness toward animals will hopefully take you to the next stage of possible relationships. Up until now, I've been discussing how to observe what is going on around you and how to act in a less disruptive manner. Taking this a step further, one of the highest compliments from our undomesticated kin is the privilege of being allowed to disappear. Disappearing in this case has little to do with H. G. Wells's *The Invisible Man* or the cloak of invisibility worn by Harry Potter. Not being seen is the privilege of being accepted to the point of being overlooked, the kind of invisibility Rick Bedsworth has achieved in his forest. This is a level of transcendent grace that happens when animals acknowledge you, take note, and then carry on with their lives—and it requires you to set your thinking cap aside. To experience this kind of association with local fauna and get as close to Rick's experience as we can, we will proceed by turning our awareness inward.

Even if your invisible state lasts for only a few seconds, it can be something you remember forever. The barrier to finding communion with nature, as it turns out, depends partly on what we do with our bodies, but has as much to do with what stirs between our ears. I touched on this briefly in the last chapter, but in this chapter, let's go deeper into the details of disappearing into your surroundings.

One of my more memorable brushes with invisibility came several years ago in the company of four black-billed magpies. They are not only among the craftiest of corvids, but also some of the wariest birds. The magpie clan around our home seldom tolerates humans within fifty yards of them. Their tail feathers are commonly all that we see disappearing into the distance. Magpies even have a special vocalization—*wik-IK!*—that they use when they are surprised on the ground at close range.

On the day in question, I decided to vary my sit spot routine. Instead of staying put, I took a bit of a walk. Calling it a walk is a bit misleading. It was more like standing still and taking a slow step every once in a while. My so-called walk covered maybe thirty feet in the span of an hour and a half or two hours. In this sort of exercise, I wasn't breaking up my silhouette using background objects or hiding in the undergrowth. I was doing the exact opposite—I was presenting myself; I wanted to be seen.

Out among the sage and rabbitbrush plants, I made every effort to feel all that my senses could absorb. I basked in breezes upon my skin and relished their touch from every direction. I let the light reveal every movement of the grasses and registered small departures from those themes—such as the moment when a grasshopper sprang into flight. I welcomed every sound that met my ears: distant ravens croaked and jetliners streamed far overhead. The sensations were integrated seamlessly and simultaneously in a liminal space that left little room for thought.

As I acknowledged these sensations, I let them touch me and imagined them passing *through* me. The result was a pleasant, almost dreamy state that was supremely relaxing. Unlike a typical daydream, where you become lost in reverie and detach from everything, this sense of close connection with everything around me was refreshingly expansive. I entered a transcendent state where my own physical boundaries seemed to blur. I became defined by, or maybe even composed *of,* my senses and surroundings.

Colors seemed to be more vivid, drafts of moving air came from many more directions than I had previously been aware of. I felt saturated with sunlight as I visualized it passing through my skin and out the other side of me. A song sparrow hovered right next to my head and dropped into a clump of grass only a few feet away. Moments later, the sparrow hopped to the ground and fed a few feet from my shoes. At that point, the bird nonchalantly picked up a wild rye seed, gnashing it with its beak like someone idly chewing a drink straw at a country diner.

I don't know how long I was in that state, but it felt good, *really* good. When my telephone alarm went off—telling me it was time to get back to emails and phone calls—it was the most jarring sound I had ever heard. As I slapped the screen to turn it off, I tried to hold on to the remaining pieces of my relaxation. Slowly, I moved up the garden steps, past the greenhouse and the garage, and back to the house. This was when I saw the magpies.

The birds didn't fly away. I just kept walking. Before I knew it, I was standing right over the birds—four of them. Had I fallen, I would have crushed the closest two magpies and probably brushed the other two with my head. They did absolutely nothing. This was as close as I had ever come to a wild magpie—and they were allowing me to approach of their own free will. What kind of mystical space had I entered?

I could see the choices the birds were making as they poked their beaks into the grass for food and how they were relating to each other on an intimate level. I was at a loss for understanding. How is it that the magpies didn't so much as look at me, much less step away? These were wild birds, not the street sweepers from town who are used to handouts. But panhandling magpies don't behave this relaxed around people—not even when a french fry is extended their way.

No one in our neighborhood has ever intentionally fed the magpies, so there was little chance of them being food-conditioned. Considering we spent very little time in one another's company, it didn't seem that they could be habituated to me in any other way either. Our non-history together made the entire episode all the more intriguing. The birds continued picking through the grass and talking to one another in low grunts and impish squeaks. I felt supremely blissful, receptive, and yes, invisible.

"I can't believe this," I mused. And the moment those outside thoughts interceded, all four magpies immediately stopped what they were doing, looked directly at me, and burst into flight. I hadn't moved a muscle. I tried approaching them immediately afterward, but they would have none of it. Thirty yards was about the limit of how near I could go before they flew even farther away. The spell had been broken.

Most of us have never known someone who has attained a heightened level of passage within the more-than-human community. Most humans alive today are standing on the outside, looking in. In nature photography and hunting circles, many think that disappearing means wearing camouflage clothing. Even with scent- and UV-blocking technology, this thin veneer is not the same as being given a free pass and accepted for who and what you are. Once the wind changes or the light shifts, you cease to be a tree or a moss-covered rock and are treated like any other threat. Camo lets you match the wallpaper, but it does not let you join the family at the dinner table.

During my hunting days—bedecked in cotton-twill tree bark and leaf-smattered coveralls—I was frustrated by my inability to approach wildlife. How could the squirrel sidle up next to the woodpecker at my grandmother's bird feeder and not scare it away, but I couldn't? I did not look quite like a human in my cryptic clothing, but I still walked like one, talked like one, and certainly smelled like one. Understandably, I got the same reception as any of my plainclothes comrades. What gives? I'm an animal too, am I not?

The boy in me who followed fox tracks in the woodlands and admired distant herons along the lakeshore, however, craved something more. At that time, commingling with a fox or a heron seemed as unlikely as landing on the moon. Perhaps you know the feeling? I eventually found that the more I changed the story I was subconsciously projecting, the closer I came to uplifting experiences. Taming the "monkey mind," as it is referred to by Buddhists—that spinning hamster wheel of inner chatter, demeaning self-talk, desires, worries, and longings—is essential. This would not be necessary if we could keep our monkey mind to ourselves, but as evidence shows, we cannot. We are continually telegraphing our inner electrical storms to the outside by countless unconscious signals. We don't seem to know we're doing it, we can't seem to help it, and animals are at the receiving end.

If there was any doubt about whether we are sending out more messages than we think, one need only look to the example of Clever Hans. Clever Hans was a Russian trotting horse who had the rare ability of doing math, telling time, and spelling words. Hans's handler carted him to festivals and public gatherings all around Germany in the late nineteenth and early twentieth centuries to display his skills. Hans was a sensation, and he drew crowds by the thousands.

During demonstrations, Clever Hans's handler would hold up a card displaying a number or math problem. Hans would tap his front hoof on the ground the number of times of the answer. Hans could do everything from simple arithmetic to basic square roots and even fractions.

With the attention lavished on this equine wunderkind, scrutiny was inevitable. In 1904, the same year that Hans made the front page of the *New York Times*, German education authorities decided to take a look. They gathered thirteen experts headed by psychologist Carl Stumpf to investigate. In addition to Stumpf, the Hans Commission, as it was known, included a horse trainer, a veterinarian, a cavalry officer, a circus manager, a number of schoolteachers, and the director of the Berlin zoological gardens. Their conclusion? Hans was legit. There was no fraud.

After the commission did its part, the matter was handed over to a psychologist by the name of Oskar Pfungst. Pfungst found that the handler—who thought he was a benign presence in the equation—was the key to Hans's abilities. As the handler was instructed to move farther away from the horse, the horse became less sure of his answers; most critically, if the handler himself did not know the answer to the problem, Hans didn't either. The success rate of the stallion under these conditions spiraled downward and finally crashed in a Hindenburg-style heap of zero. Pfungst's revelation sent shock waves through the animal-behavior world. The "Clever Hans effect," as it was dubbed, was real—the observer can seriously affect the behavior of the observed without ever realizing it. When it came to animal intelligence, this seemed like an epic fail. The horse couldn't do math and he couldn't tell time or dates, after all.

The Clever Hans effect, however, betrays a different sort of genius, in my opinion. As opposed to an animal picking up clues in the broader environment—such as a dog sniffing out a buried corpse or a snake using infrared receptors to find prey—here was a willful act of precise decision-making on the part of a nonhuman, a horse in this case, based on subconscious cues from a nearby human. Even as we swear that we are not externalizing our thoughts and desires, we are indeed sending out more information than we think. We continually broadcast our inner world to attentive domestic and wild nonhumans, and they are clearly getting the message.

Geoffroy Delorme, author of *Deer Man*, found that moment-to-moment status reports billowed out of his skin and into the nostrils of his wild roe-deer companions. The odors seemed to convey everything the deer needed to know about his mental, physical, and emotional state. The cloud of odor enveloping each of us varies greatly when we are relaxed versus stressed or scared, but have you ever tried to prevent those compounds from coming out of your pores? Although we can lie through our teeth, it is virtually impossible to lie through our skin. The deer sensed it, interpreted it, and responded in kind, just as the wolves at Mission Wolf did with Shauna and her fellow workers. The things we spell out in the absence of words can be quite disturbing to sensitive people and animals alike. Wild animals, in particular, have a knack for picking out whether someone is trustworthy and can be safely ignored, or whether they need to remain vigilant. Why? Again, their lives depend on it.

To put it another way, animals, like most people, don't like to hang around with people who make them feel uncomfortable. I think you know exactly what I mean. Even slightly awkward social behavior is enough to sway your choice of companionship. If a person directs too much eye contact your way or a coworker doesn't pick up your overt signals that they are crowding your personal space, you often give them a wide berth later on. If in doubt, you might follow the example of dedicated pet owners and let your cat or dog gauge the company. If a guest gets the cat's dander up, maybe you don't want anything to do with them either.

We intuitively grasp subtle messages layered into the voices and movements that other people use. We immediately recognize "the look" from a parent or a displeased roommate. No words are needed. Similarly, think of when your significant other steps through the door after a long day. It's obvious whether they have big news to share or whether it has been a rough day on the job. There's no need to talk about it. You intuit that it's your turn to cook dinner and that serving a heaping bowl of ice cream with extra chocolate syrup for dessert is in order. The message is clear.

So sensitive are animals to our intentions that it is codified in abundant Indigenous lore. There are beliefs held among Athabascan, Apache, and other peoples that one should not state an intention to go hunting out loud. Rather, if anything is to be said, tangential references should be made. Telling someone you are "going for a walk with a full quiver of arrows" might be a good work-around. This nugget of wisdom recognizes the subtle but important details of subconsciously embodied intention.

So how do we reorganize our involuntary news feeds? One way to do this is by using what I call the bridge to silence. When I teach classes for educators, I sometimes take them to a narrow suspension bridge that crosses the Gardner River in Yellowstone. I give only a few simple instructions, such as to remove all watches, phones, or any other timekeeping devices, and put away all pens, paper, and notebooks. I then inform them that once we cross the bridge, there will be absolutely no talking permitted until we return.

After the students cross the swaying, bobbing bridge and have selected a seat, I set my stopwatch for exactly sixty minutes. Unlike the use of a typical sit spot to take in what is going on around you, the only guidance for this exercise is to spread out and clear your mind. When the sixty minutes are up, I signal with a hand gesture and we cross back to the other side of the bridge.

It is always interesting to observe how long it often takes people to begin to talk again. Even after returning to the original side of the bridge, people are hesitant to speak. Silence brings with it an air of sacredness that many seem reluctant to violate.

"How long do you think you were sitting out there?" is always my first question.

Forty-five minutes, seventy minutes, an hour and a half are typical estimates. Most feel that it was much more or less than the actual sixty minutes. When I tell them it was exactly an hour, many can't believe it. They wrinkle their brows and glance about in confusion. Time gets distorted.

My next question gets to the heart of the lesson. "What did you find during your time?"

A student might share details of a bird landing in the tree above them. Another might notice snow fleas in the dirt. Some may discover the exquisite geometry of a pine cone or the overlapping scales of juniper foliage. Still others, somewhat sheepishly, admit that they just fell asleep.

"This is okay," I tell them. "Don't worry about the snooze. That's what your body needed to do."

"This was *really* hard," one of the students might say. "I have never done anything like that before in my life—being told to sit down, not to do anything, and try not to think about anything."

"I wanted to get up and walk around *so bad*," someone else invariably adds.

Some discover a soundtrack of the same few songs playing on a loop inside their heads the entire time.

"I loved it," others will say. "How often are you told in a class to go out and sit somewhere and not to do anything?" Or "It was great! That was one of the best assignments *ever*! I got more relaxation out of that hour than I've had in days!"

All seem to agree that despite the difficulties with sitting still, the experience is eye-opening. Many allow that they would like to do that more. In the face of their initial challenges, most students agree that it feels great to give in and quash the compulsion to do something. From my perspective, this is the real value of the exercise. The bridge to silence is where we start to dump some of our baggage.

As we have seen, quieting the mind is no small task. If you have ever tried meditation, you have likely found that sitting still and not thinking about anything can be a Herculean task. This is where nature helps us in special ways. Each time your mind starts to wander or the desire to get up and go for a power walk emerges, recline back to your senses, nothing more. Unlike previous exercises where we looked deeper into things as a way to extract information or piece together relationships, here we are using the beauty and variety of nature to pull us out of our thinking, doing habits.

Staying in the moment can be difficult, but keep at it. As you absorb the outward sensations, give equal weight to inner sensations. Note them in a detached manner, the way an outside observer might, then let them go. Are you feeling stiffness in your neck? Take a lead from the breeze that is blowing and imagine it gusting right on through those areas of stress and tension. Let it all go. Did you have a frustrating conversation earlier in the day? If so, allow those shimmering leaves in the treetops to arrest your thoughts and attention, and then go up into them using your mind's eye. Feel what it is like to lilt from side to side with thousands of other jostling leaves around you. Acknowledge the troubling news you received from your doctor and place it under a stone. Picture the microbes in soil organisms chewing up your worries and fertilizing the ground for better things to come. Be where you are; be here. Be in the moment.

Where will this kind of experience take you? This part of the journey is meant to take you deeper into the recesses of your own human nature. Letting go serves to attenuate the divide between our inner and outer spheres and brings them closer together—this is where you will find your sincerest comfort and greatest strength. This is the pace and wavelength that the rest of nature is operating on. Mother Nature is great at offering a system-wide reset so that, once in the zone, we don't have to worry as much about committing disruptive acts, such as breaking a stick when we are out walking and hoping to see wildlife. After all, turkeys and deer break sticks all the time and no one runs.

Repeated resets of our inner landscape automatically re-organize the messages we project to animals. A state of calm and relaxation also predisposes us to detecting far more subtle cues from our surroundings, and this goes a long way toward endearing us to nonhumans. Next time you feel the itch to scroll through social media, try putting the device aside and crossing the bridge to silence instead.

Up until now, a lot of the emphasis has been on things to observe, stuff to know, and things to do. The bridge to silence is about setting all of that aside—it's about *being* rather than doing. When crossing an actual bridge, a threshold, or any demarcation, create your own routine that turns the lens back upon you. Do I feel groggy or calm? Am I rested and alert? Follow the joy in things, pursue those activities that offer you solace and inspi-ration. Consider using your existing sit spot differently, take a prolonged walk over a short distance, or go snorkeling in a less directed way. Replace exploration in favor of floating along with the swells. You will be better off for going with the flow, and your animal neighbors will be some of the first to notice the difference.

Although you might find it difficult to clear your mind at first, no matter what degree of success you find, you will be leagues ahead of where your cluttered mind was before. Make room for the expansion of inner silence and the elevation of a lighter, brighter you. And when the soundtrack starts up again, or another item on the to-do list leaps to the fore, acknowledge it, set it aside, and refocus your attention on something in the present. Breathe slowly in through your nose, filling your belly and then lungs, and exhale through your mouth. Repeat this breathing several times until you feel yourself unwind—hold this space sacred. Pulling in and letting go in this way releases you from the confines of the clamoring voices in your head and all those counterproductive subconscious messages you are sending out into the world. You are creating a refuge from the pressures to get things done, to be something you are not, or to flit from one disjointed thing to another. You are giving yourself the gift of time.

Once you cross your own bridge to silence, place no expectations on the agenda. And if you do anything in association with this practice, let it be setting a date with yourself to do this activity on a recurring basis; put it on the calendar and defend it like you would a long-awaited vacation or a trip to the doctor's office. The author of *The Artist's Way*, Julia Cameron, recommends establishing "artist dates" to plan for unstructured time. Allow a timeless state of suspension to rule. If we really want to work on our relationships with nonhuman beings, it starts with working on ourselves. Once you are on the other side of the bridge, you will radiate a quieter, more tranquil state, and when your inner and outer spheres balance with that of nature, you will find greater communion with all things.

Rick Bedsworth told me that when he has "arrived," he becomes incredibly relaxed, his breathing slows, his thinking mind is quiet, and he is filled with a sense of expansive awareness. Sometimes he will pick a landmark, a tree for instance, and say to himself, "By the time I get there, I will be in my quiet mind and relaxed in my body." Other times, he will count down from an arbitrary number, and by the time he gets to zero, he is in the zone. Experiment with what works best for you. Try starting a nature journal as an entry point if sitting still cold-turkey for a long time is a challenge. When the time is right, instead of feeling the need to be doing something, you will be able to set the journal aside and just be. Experiment with your own versions of these helpful aids. Be gentle with yourself.

What we find in our wild companions is a reflection of ourselves bouncing back at us. The response we get from these untamed counselors is a gentle reminder to relax, drop our troubles, jettison the ego, and find our own better nature. What Rick and others describe most closely resembles the state of practiced meditators. In my own case, I have gotten this feeling only in small doses, but it is intensely powerful. Once you find your way to this special ground, you will want to seek it again and again.

- 20 -

THE POWER
OF ONE

BUILDING RELATIONSHIPS with individual nonhumans has a way of shifting the ground beneath our feet. Those of us who have pets or service animals know that the bonds we build with them can sometimes exceed the connections we have with other humans. Knowing someone beyond the species divide can underpin scientific advancements and trigger personal transformations. In primatology, Jane Goodall and Dian Fossey introduced us to personalities like Flint, Gremlin, Digit, and Kweli, and as a result gave us an insider's view of never-before-observed behaviors (or at least never observed by Western scientists). The chimpanzee David Greybeard, for instance, was the first to demonstrate tool use to Jane Goodall. Along the way, these women became some of those animals' most dedicated advocates.

In his book *A Primate's Memoir*, neurobiologist Robert Sapolsky echoes the importance of getting to know individuals. Sapolsky spent twenty-one years studying savanna baboons in Kenya. Specific baboons brought him into the fold; he witnessed their joy, trickery, shifting alliances, and simmering grudges. Sapolsky mentions a conundrum he faced when visiting other researchers and their baboons. He writes that he felt "something almost like irritation, [because] I don't know them, I don't know who they are. How can I appreciate them without yet knowing who is who? What great fights are going on and who are the epic personalities? It's like being a few pages into some epic novel and getting frustrated... [Which] one of you is about to have a life history that change[s] all my ideas about baboons?"

Species more distant than our primate kin can also wow us with the power of their individuality. An eight-year investigation of wild mule deer near Lander, Wyoming, became a singular obsession for Joe Hutto, whom we met earlier on the subject of wild turkeys. Joe could count the number of days he'd been away from those deer on one hand. A chance encounter with a single buck convinced him that "this wild animal was able to see me as an individual and that I granted him his individuality... I was not seeing some *thing*, I was seeing some *one*." Joe had many breakthrough experiences with the mule deer, and one of them, a doe he named RagTag, bestowed on him a remarkable honor— she took him to her sanctum sanctorum, the hiding place of her newborn fawns.

Naming animals still comes with stigma in scientific circles, but doing so helped Joe identify over two hundred individual deer. None of the deer had ear tags, radio collars, or any other human-made markings. Quite astonishingly, many of those deer also responded to the names Joe had given them. He could call an individual deer's name and it alone, out of a herd of dozens, would walk over to him—not unlike a pet dog beckoned at a park. This is a surprising trait in an animal with few or no coevolved bonds with humans. How many other species might show similar traits if we give them a chance?

Starting a relationship with a wild animal can be as easy as noticing a missing fin, an oddly colored feather, a limp, a scar, or an interesting pair of antlers. Friends Joe and Natalie in Maine came to know two young tom turkeys they called Waldo and Willy, based on their unique beards—a modified bundle of feathers that looks like a black paintbrush projecting from the middle of a male turkey's chest. Waldo and Willy were lower on the pecking order in their flock, and when they visited the bird feeder, Joe afforded them a measure of protection against the more dominant toms. The relationship went two ways: Joe learned about the intricacies of turkey hierarchies and temperament, and Waldo and Willy knew Joe had their back.

Everyday life doesn't offer most of us the luxury of spending entire days or months, much less years, focused on the lives of other species. This doesn't mean we can't bask in the glow of repeated encounters, though. Ambassadors of the animal kingdom come in many shapes and sizes, and can be found in abandoned lots, town parks, and the vegetated margins of any street. When a particular ambassador from across the divide does catch our eye, they have a bewitching way of pulling us deeper into their lives whether we intend to go down that road or not.

For Guy Galante, my group leader in Jon Young's bird program, what started with the goal of getting a photograph turned into a fifteen-year cross-species journey. The setting of Guy's story was a tiny strip of greenery bordering the American River in Sacramento, California. With nearly one million people calling this landscape home, the American River Parkway sees more human visitors than Yosemite National Park—which has an estimated eight million user-days each year. This unlikely wilderness serves as a haven for a wide array of wildlife and provided an opportunity for Guy to form new relationships. As the parkway was a mere fifteen-minute walk from the public school where he taught, and not far from his home, Guy spent a lot of time there. In a nostalgic look back, he says, "I could do it all blindfolded now, but then it seemed so vast."

Digital photography was just becoming popular back in the early 2000s, when Guy started a project inspired by *The Outdoor World of the Sacramento Region* by Jo Smith. The book was illustrated with black-and-white drawings of area organisms, and Guy set out to photograph every subject. Armed with his 3.2-megapixel Olympus point-and-shoot camera, he captured images of plants, eagles, and fungi. Within a few years, he had taken photos of most of the mammals too. But there was one conspicuous absence in his portfolio—the coyote.

The challenge of catching a shot of the elusive canid fueled Guy's imagination, but he just couldn't manage to see one. Other parkgoers might report, "Yeah, I just saw a coyote over there" or "I've been seeing one right around here within the last few days." Guy was always a few steps behind. It almost seemed like the coyotes were deliberately ghosting him.

On September 4, 2005, Guy finally got his long-anticipated picture of a coyote—a handsomely coated individual running through the frame from right to left, with three feet suspended in midair. Getting that shot felt like "no feeling I ever had before," Guy tells me. He had been focusing on this goal for so long that the brush wolf had become elevated in his mind to an almost mythic status. He was hooked. From that point on, the thrill of capturing a photograph of a coyote became Galante's central focus, and along the way, Guy found a friend.

Guy dedicated all of 2008 to figuring out how to capture more urban coyotes on camera. Naturally, his camera got much bigger, and fifty thousand photographs later, he has caught the attention of many more passersby.

"Are you photographing birds?" they often asked.

"Actually, I'm photographing coyotes."

"Oh, there's coyotes around here?" people would respond in a tone often laced with no small amount of surprise and trepidation.

I ask Guy how he handles interactions like this, and what the most important things are that the public should know about coyotes.

"Coyotes don't eat people" is Guy's snap response.

There have been only two coyote-caused human fatalities documented, ever—one of a three-year-old girl in Glendale, California, in 1981 and another of a nineteen-year-old woman in Nova Scotia, Canada, in 1994. Roughly 4.7 million dog attacks take place each year in the United States, by comparison. And between 2005 and 2019, 521 Americans were killed by domestic dogs. You are far more likely to be assailed by a dog-walker's pooch than a coyote, by a long shot.

The more Guy acquainted himself with the American River coyotes, the more he understood that they were not a revolving cast of anonymous characters, but many of the same individuals popping up over and over again. Familiarity with them grew to the point where he had yet another epiphany: these were not transient coyotes—they were *residents*. They lived here, raised their kids here, and dodged the same people and domestic dogs day after day, year after year.

Identifying the residents made Guy feel like an intruder at first, but in time, it led to great strides in his understanding of and appreciation for these wild canids. He found himself wanting to know what their lives were all about. "I knew," he tells me, "that if I kept going to this area, I was going to cross paths with them in this very narrow strip of the parkway." Late in the summer of 2015, Guy met a member of a local pack, and their encounter threw the doors wide open on the lives of American River coyotes.

Guy saw a smallish coyote interacting with a litter of puppies. This subadult was playful and attentive to the youngsters, but most strikingly, he had only one good eye. Although the left socket contained an eyeball, it was black and wept profusely, the result of some unknown injury. Guy dubbed him One-Eyed Jack, or OEJ for short. This is where their relationship began.

Jack's injured eye and habit of laying his ears out somewhat flat made him stand out among the small crowd of urban coyotes. Perhaps out of a desire to root for the underdog, Guy was smitten. He and Jack would share many special moments over the next two and a half years. Although Guy became familiar with many of the other coyotes on the scene, Jack was the first and paved the way for unforgettable days with Sacramento's urban wild dogs.

Guy chronicled his time with One-Eyed Jack with his photography and on social media. The following are a few entries from his Facebook page, "Roots of Connection":

July 6, 2017: "OEJ cautiously but confidently strutted his way across the park. Haven't seen him in a while. I'd stopped on the access road to take his picture. Some dude (in the only other car on the road in the entire park) pulled up beside me and yelled, 'Lash your dog!' And yes he said lash. And then he sped away." (In the end, although he obviously never tried to put a leash on Jack, or to feed him, Jack became, in a way, "Guy's dog.")

April 7, 2018: "There was a 5–7 minute period where it was just 'One-Eyed Jack', [another coyote named] Natoma, and me. Pure bliss. There was no one around, the winds were powerful, mine was the only car in the lot, and the three of us were together. It felt good to be in the presence of OEJ."

May 15, 2018: "OEJ and I scared each other to death today! We both had a small ring of disturbance but also a small ring of awareness. I was sneaking up to a spot where I might see the coyote pups [in Jack's family] and in the process I literally walked within ten feet of a sleeping OEJ who awoke and jumped up. While it was thrilling to be so close to him, I felt so bad and apologized to him profusely! It appears he has a new boo-boo on his nose."

June 29, 2018: "OEJ had moved NE out of sight, into the thicket. So, I turned my attention to the SW where I thought I might see the pups. I hadn't traveled but 100 yards from the OEJ sighting. Then I hear a panicked call behind me, 'Booker! Booker!' The woman was frantically calling to her dog. I turn and see Booker, a yellow lab, chasing OEJ! And he caught him! Holy shit! ... I have over 35 interactions documented with OEJ. He's pretty chill. Never known to be aggressive." (An image of the encounter shows One-Eyed Jack facing off against Booker. Jack has his mouth wide open, tail tucked, and body hunched in a defensive posture.)

August 28, 2018: "I feel extremely lucky tonight, and also a little worried. I have never photographed a family group all together like this! I am rethinking OEJ's role in the family—he seems to be high up in the hierarchy—[I see] him standing tall in the middle of the engagement! The grown-ups are exhibiting behavior that indicates it is time for the juveniles to move on. The juveniles will soon start to find their own space and this often means they'll wander into suburbia more—I wish I could tell them not to do it."

September 7, 2018: "My heart sank when I turned around and saw what I thought was a dead coyote only fifteen feet behind me. I mean, a sleeping coyote would have woken up with me that close right? I gulped and took a step off the trail to go identify the corpse. But... it was OEJ! He sprang to his feet, gruffed at me, as if to say, 'You woke me up, man!' I said the same thing as he brought me to presence. He retreated a few feet, and promptly sat down and scratched himself. Two things: maybe he doesn't hear so well? And his left eye is looking really good!"

January 16, 2019: "So happy to see OEJ alive and (mostly) well today!" (His bad eye would go through different phases of being very weepy and not looking very healthy at all, and in other cases both eyes wept, the good and the bad.)

February 1, 2019: "OEJ seems very loyal to his partner who has very poor eyesight. He stays very close and seems to be protecting him/her. I wondered if s/he could see at all, but s/he can—barely. Jack looks like he's in bad shape too, but much improved from six weeks ago. I liked watching them interact, but it hurts my heart at the same time."

February 9, 2019: "I'm in love with this photo of OEJ and his female companion walking together, stride for stride. They're perfect together! As touching as the image is, the animals look a bit roughshod, mangy perhaps on account of the rat poison that has accumulated in their systems." (Over-the-counter anticoagulant rodenticides are a serious problem for predators worldwide.)

Guy would get many more photos of the American River coyotes. The dopamine hit from his encounters made him wonder at times if he was a little bit nuts. Was he using the coyotes to fill some void within himself? But time brought the matter into focus. Being in the coyotes' company made him supremely happy. Spending time in this urban jungle with OEJ and the gang felt good to Guy; it felt right.

In an article by a reporter for *Sacramento Magazine*, Guy shared stories of some of the other individuals he had come to know. The reporter wrote of "Bolt, with electric green eyes, who bolts the moment she sees a human; her dad, Tux, with a white chest, who's more relaxed; and her mom, Notch, who's a watcher of all things; The Howler, who howls every time he sees a dog; [and] Mama Bear, who's always on the move." The reporter noted that Guy talked of the coyotes' "playfulness [...] their athletic grace, and their tender licking of each other's faces. They come to life as he speaks and I scroll through their photos."

During our conversations, Guy speaks affectionately of these coyotes. "People that I meet in person that were just following me on Facebook would ask me about OEJ like he was one of my family, which I kind of guess he was."

We come to see not personhood in individual animals—this would be a supreme anthropomorphic distortion—but something we could more easily call personality. Is the sense of seeing and being seen part of what attracts us to those others? Many feel that this need to connect with nonhuman beings is rooted in some level of "species loneliness," as Richard Louv calls it. E. O. Wilson warned of similar ills and said that if we don't curtail the mounting dangers of the Anthropocene, we could be headed for the Eremocene, the Age of Loneliness. We have come so far, but we have also become quite isolated. The social animal within us screams for companionship, human or nonhuman, and when we do find a special connection, the more we give, the more we get.

"Knowing an animal at a deeper level does not diminish its mystery," writes Louv. "[It] opens our hearts to the unknowable and, in a sense, the divine." People who have found close, sincere connections with the broader community of life don't let go of these connections easily. It takes only one individual with a blind eye or a tattered wing to welcome us into a bigger family. After taking that first step, we can't go back. Never again can we look upon a member of that animal race and not afford them the same potential for connection as our close acquaintances.

There were times when Guy would go to the park and simply sit in the company of Jack and the others, without taking pictures. A relationship had clearly developed, and it seemed to go both ways. On occasion, a hiker or biker might come down the trail and Jack would disappear. Once the stranger had passed, the coyote would return. Guy was no longer a stranger to the American River coyotes; they trusted him.

Relationships of all kinds would come and go, but Guy's connection with the American River coyotes would endure. Despite the hardships they have suffered as a result of attacks by domestic dogs, collisions with automobiles, and encroachment from people, the coyotes persist. They remain good parents, they protect their homes, they stand by their partners, and they show us that there is always hope.

Populate your wilder community with familiar faces. Pick an animal that is nonmigratory. Focus on species that don't have overly large home ranges and are visible regularly—even if it requires a bit of work, as it did in Guy's case. Be aware that birds' feathers change as they mature, mammals molt, snakes shed their skins. Look for marks that last, particularly at the beginning of your relationship. Scars, unique facial features, or other distinctive markers will serve you well across the arc of time. Subtler identification marks will automatically jump out as your time together grows.

You may decide to use something like Google Earth, the way Guy did, to document the places where you see those individuals. One-Eyed Jack, as it turned out, got around far more than Guy realized at first. Guy would sometimes run into him in unexpected places or learn of Jack's wanderings from the sightings of other people, and there were other times when Jack was nowhere to be found. It made Guy wonder, "Where did you go?" You may be surprised by how much your nonhuman familiars get around and what lengths they must go to in order to survive.

Take the long view of your budding relationships, as they can take time to cultivate. All the best ones do. Relationships are all about mutual respect. Don't expect to tromp off into the wilderness and commune with the animals without doing a lot of homework beforehand. To kindle connections with wild nature, we must be willing to give of ourselves. Giving from the human perspective may be as simple as not minding the inconvenience of offering animals a wide berth when on a walk, lending your attention to their preferences and idiosyncrasies, or investing the time required to get to know one another. It's about seeing them for who they are, not just as the object at the other end of a camera or gun. It's not about collecting sightings or trophies; it's about relationships. Weeks, months, and even years of deliberate, gentle introductions may be needed for small bits of headway. Invest whatever amount of your precious time you have, and you will be more than rewarded.

As transcendent as becoming attached to individual animals can be, it can also be heartbreaking. Watching your friends suffer and even die may spur you to want to intervene. Many wildlife advocates have been moved to action by these convictions. Stepping in, though, can often backfire. "Lost" baby birds are usually fed by their parents for weeks after leaving the nest. Fawns stashed in flower beds or near back doors are seldom abandoned by their mothers, and deer that look as though they are starving in harsh winter months can be killed by supplemental food that their bodies are ill equipped to handle. The latter have dialed down their metabolisms to make it through the hard times, and giving them extra food is doing them no favors. Our initial instincts are guided by our highest intentions, but in all but the most extreme cases, it is best to have a hands-off policy.

Our family knows these struggles all too well after getting to know a beautiful female mule deer we called Momma Deer. In her last year, Momma Deer had three spritely fawns. One by one, the darling fawns died with the onset of a harsh, icy winter. The urge to feed Momma was overpowering. When we learned more about deer survival in winter, we refrained. Not only is feeding deer illegal in the state of Montana, but getting the wrong nutrient combinations can be damaging. And if we were successful in helping the deer survive when they would not have done so without our intervention, more deer would likely open the door to disease outbreaks or start to attract large predators such as cougars and gray wolves, and we would be building a larger and larger population of deer who would be dependent on our handouts. If there came a point where our budget didn't allow for the increasing cost of food, what might happen to all those hungry deer? What if we moved and the next owners didn't maintain the feeding? There were too many confounding variables.

In December, Momma also disappeared. We feared that she was gone for good. We were stunned when she returned in late February. Momma was scarcely recognizable in her pitiful condition—she was little more than a bag of bones. As crushing as this situation was in those final days, Momma bestowed on us one of the highest honors. Before her inner fire flickered out, she bedded down on our welcome mat.

In her worst hour, when any predator could have easily overtaken her and no food could stem the tide of her decline, Momma Deer remembered our home and sought its safety. Our ties to Momma were built on trust, not handouts. For two evenings, Momma bedded on the doormat on the raised decking of our porch. I sat by her side on the floor separated only by a pane of glass. Tears rolled down my face as I watched her shallow breathing beneath her emaciated form. Under our protection, she bedded beneath the deck on the third night and expired.

If you make space for an individual creature to reveal itself to you, be prepared for many unexpected outcomes. Commit from the outset to focus on what's best for your newfound friends, even if that means avoiding interaction or close physical proximity. Building a relationship with someone of another kind is a privilege and it comes with substantial responsibilities. If it ever looks like your time together will do them harm in any way, give them their space. Few things are as difficult to earn, and as easy to lose, as trust. And if you find yourself brought to your knees by the fate of your nonhuman family, my sincerest condolences.

Know this: the pain you experience is only a reflection of the beauty you found together.

July 12, 2019, would be the last day Guy would ever see One-Eyed Jack. No one knows what happened to him or where he finally came to rest, but what he gave to Guy Galante remains to this day. Poster-sized prints of Guy's cherished American River coyotes grace his walls, and the one positioned for easiest viewing in his living room is, of course, of Jack.

− 21 −

MAKING IT
YOUR OWN

OPENING THE DOOR to wilder experiences shows how much rests in our hands. Supporting our wild friends and teachers also means taking care of the land. Each one of us is the shepherd of a given place and, therefore, a custodian of nature's conversations.

Gloria Garrett made the decision to move back to a landscape that had tugged at her heart since childhood—the Leelanau Peninsula of Michigan. Gloria has a commanding yet inviting presence and says of her upbringing, "My parents were having a hard time when I was born and they weren't able to really be there for me." Outdoor spaces and their inhabitants became a source of companionship and solace for Gloria. When she reflects on the best of times, what springs to mind are family vacations in Michigan.

Thinking back on those days, Gloria pictures "sitting on a blanket on the hillside, looking down toward the little lake while my mom painted happily. It was spring and still a bit chilly. The sun was just coming up from behind my mother and the hillside was covered in dew. Fog was drifting off the cold water and creeping through trees. Deer, rabbits, and a raccoon appeared and disappeared in the mist." In this space, Gloria tells me, she felt "connected to everything—everything I needed was in that moment." She pauses and then says, "There is no separation between who we are and what is going on around us."

After Gloria and I first became acquainted at a workshop in Yellowstone, she made the leap from California to Michigan on the heels of some COVID-induced soul-searching. Her wish was to be *of* a place, rather than simply from an address—she wanted to feel that wholeness again. As the late poet and artist-philosopher Tom Jay articulated, we are meant to be localized creatures; we hunger for it, and "the first obligation of the new community... is to stop moving." It seemed that Gloria had found her way back to the place that made her feel whole.

These days, we jet around, globe-trotting among any number of destinations, seemingly delighted by being untethered to one geographic location. The result of our kinetic lifestyle, however, is that we do not know the intrinsic goodness of our homes. By not staying put, our uprooted ways also lead us to commit countless trespasses against the world that we would never consider otherwise.

But won't we get bored dedicating so much time and energy to one place? Conger Neblett Hagar knew the value of staying put and that boredom was not to be feared. Known to friends as Connie, Hagar was visited by author Edwin Way Teale in the mid-1960s for his series of books *The American Seasons*, one of which won a Pulitzer Prize. "Twice a day [for thirty-five years], storm or fair weather," he wrote, Connie "patrolled the same four-by-seven-mile area beside Aransas Bay ... Her discoveries have attracted thousands of ornithologists and advanced birdwatchers to her small section of the central Gulf Coast."

Connie was not a scientist; she was someone who rejoiced in the splendor of what surrounded her. She didn't just look at birds and record their names, she got to know them—and know them she did. Connie knew birds not only by their looks, but also by their actions, attitudes, and personalities. Along the way, Connie added more than twenty species to the Texas state bird list. She revealed previously unknown migratory patterns and range distributions. And in a three-day stretch in spring 1952, less than a mile from her home, she observed one of the last Eskimo curlews—a bird largely considered to be extinct at that time.

This petite, unassuming woman helped put South Texas on the map as a bird-watching destination. And when she was asked if she might like to travel and visit other places and exotic locations, she said she would do so as soon as she knew all there was to know about Rockport, Texas. Two wildlife sanctuaries would later be named in her honor.

Home does not have to be a place by a lake or a coastal flyway; it is found wherever our deepest attachments reside. As traditional cultures have told us, the land provides all we need. Even in urban settings, making the most of precious moments can help us feel close to a place. Tuning in to a landscape, through all the ways it could be speaking to us, is not only a way to survive, but also a means to prosper. Aside from making the commitment to stay put, how can we start the process of redefining our relationship with home?

Gloria Garrett was trying her best to embrace her new surroundings. She started by following the path of reciprocity espoused by one of her heroes, the author, scientist, mother, and member of the Potawatomi Nation Robin Wall Kimmerer. But Gloria was experiencing a particular struggle that precipitated one of our phone calls.

"I've made this bird smorgasbord with all sorts of feeders and food," Gloria tells me, "but it's been almost three weeks and no one has shown up. What should I do?" I ask her whether there are any squirrels around. I was thinking of how the gossips have a way of sharing good news as well as bad. When Gloria confirms she has both squirrels and chipmunks in the neighborhood, I suggest that she put food out specifically for the squirrels as a treat to get the woodland conversation rolling. Feeding birds is quite common—an estimated fifty-nine million people do so in the United States—and many people derive a great deal of plea-sure from this activity. I wanted to help Gloria attract birds to the place she now called home. As with feeding deer, however, there are costs and benefits to weigh. Bird feeders can help birds survive harsh winters and long migrations, but they can also be a focal point of unintended consequences. Some will avoid feed-ing birds so that they don't spread avian pox, risk window strikes, give the neighbor's cat something to snack on, or attract non-target species such as rats, raccoons, or bears. It is always best to evaluate the choice to feed on a case-by-case basis and, ideally, pair it with improvements to local habitat and wild forage.

If the previous owners of a residence did not feed the birds, it is not unusual that the amenities offered by new homeown-ers are missed; they are not on the map. Following my advice, and going the extra mile to show her commitment to spoil her neighborhood squirrels and beckon the birds, Gloria purchased organic, sugar-free peanut butter and began smearing it onto the bark of area trees. As she coated the boles of birches and hemlocks, she did so in an outwardly spiraling pattern, reflecting the manner in which she hoped news of her birdy bounty would grow—and it worked.

The squirrels did their job of sharing the news—within two days, the first chickadee arrived. A day later, there were two chickadees. The bird traffic increased. Soon, tufted titmice, house finches, red- and white-breasted nuthatches, American goldfinches, northern cardinals, and American crows joined the flock. The extended form of family that Gloria gained nurtured her spirit and motivated her to find other ways in which to orchestrate a "co-creative landscape," as she calls it. One of the ways she did this was by naming things.

In addition to small deeds, our use of language—designating names, in this case—is a great way to start reinhabiting home. Inside our houses and apartments, we have discrete spaces we refer to as the living room, the kitchen, or the entryway. These are our most familiar haunts. Try expanding this practice outward. Don't overthink it; let the names *occur* to you.

In the first years after moving to our home on Yellowstone's northern boundary, Jenny and I found that we lacked the vocabulary to relate the details of this place. One of us might find the location of a new nest or the bedding spot of an elk, but we found it hard to describe. A sense of need led us to create our own labels for the wider surroundings. There were places and features that became known as Sitting Rock and Notch Loop. Buffalo Hill is an elevated knoll near our mailbox where many of the park's wild bison winter. We named a shallow north-south hollow between two gentle hills just north of our house the Draw. We also have our Winnie the Pooh Forest, where young George did some of his first climbing and exploring expeditions along the creek as a toddler.

There is a spot we call the Sneaky Forest, a dense grove of aspen that is cool and refreshing even on the hottest days of summer, and of course, there is the Tree Palace, a stately group of cottonwoods next to our house where Jenny has her sit spot. None of these labels have shown up, or ever will show up, on a road atlas or public map; they are ours and ours alone. Countless wild dramas have unfolded in and around these spots over the years, and these pet names help us speak to those events. The designations continue to expand to this day as we explore ever deeper.

This practice of labeling discrete segments of places is not unlike the concept of "home water" espoused by dedicated fly-fishing folks. In an article for *Hatch*, Todd Tanner writes that "the whole idea of home water comes down to intimacy, to the belief that we can grow to know and love a particular lake or stream in a way that few, if any, others can." I think this description would resonate with plenty of anglers from the past and present. What are the flow patterns and timing of insect hatches, where are the deepest pools, and how do these change across the seasons and years?

This attachment to a location is essentially what the famed British naturalist Gilbert White celebrated in the late eighteenth century and chronicled in his *Natural History of Selborne*. Recasting our relationship with place helps us listen in to the details and pay homage to the hidden darlings just beyond the doorstep. Develop your own master map of special places. Maybe dedicate an entire page or the inside cover of your journal to this practice. Call out spots that speak to you. As contextual as animal communication can be, finer scales of understanding will only improve your ability to interpret what is going on. Observing more of what animals say and do in these settings will draw you further along your journey. Have fun with these names. Start applying them to your usual haunts and share them only with your closest friends and family. Involve your kids or grandkids, nieces and nephews, or the neighborhood children. This is a task well suited to creative young minds, who will often come up with things you would never have considered.

You may already know of a particular log that is a favorite among sunbathing turtles. Dub this log the Carapace Canteen or simply the Turtle Log. Maybe there is a particular power pole that area birds flock to and use repeatedly—why not label it the Perch Pole or the Lookout? Whatever you do, make these names descriptive or funny, and anchor them not to historical events and distant characters, but to your personal experiences. Jenny and I sometimes get confused looks if we slip and mention "our" names in the company of others—they have absolutely no clue what we are talking about.

Invoking a place-based vocabulary helps to reestablish the living, connective tissue to home ground. Paint a clearer picture of recent events by invoking personal landmarks, such as when a hawk flies from Cool Spring Hollow into the north side of Pigeon Park, then flushes a group of starlings out of Eden before taking off toward the highway overpass. You will remember details and events better and describe them more accurately. The familiarity will improve your ability to interpret what you find there, but take your time, be deliberate. Let area relationships guide the process as these names become inextricably woven into your most precious memories. And as those challenging moments in life inevitably come, draw upon one of "your" places in your mind and go there.

Your special place does not need to be a destination in the conventional sense either. It might be one specific point. In some cases, a unique outcrop of bedrock might fit the bill. It could be the view from a well-placed window where you take your viewing or listening breaks. It could be as simple as a particular tree. In her book *Rewilding the Urban Soul*, Claire Dunn notes that the city of Melbourne, Australia, gave each of its seventy thousand urban trees an ID number and an email address. Although the addresses were there so citizens could report tree conditions, hazards, or problems related to vandalism, what the city received instead were "thousands of expressions of loving fan mail, existential questions, and witty tree banter... 'Hi tree 1022794, how's it going? I walk past you each day at uni, it's really great to see you out in the sun now that the scaffolding is down around Building 100. I hope it all goes well with the photosynthesis. All the best.'"

Along the shoreline of Lake Michigan where Gloria stores her kayaks, there's an ancient hemlock with a damaged base that she simply refers to as Hemlock. Although there are multiple examples of *Tsuga canadensis*—the eastern hemlock—on her property, there is only one Hemlock in Gloria's mind. Beneath Hemlock, Gloria has placed a half-circle of bearpaw-sized Petoskey stones—fossilized, whitish coral from the warm, shallow seas that covered Michigan 350 million years ago. There is a small city of Petoskey farther along the shores of Lake Michigan, but the small half-circle of stones on her property constitutes *Gloria*'s Petoskey.

Increased familiarity with a point or place exposes change, both good and bad, that takes place in our homes with the passage of time. Author Todd Tanner wrote of the downstream effects on his first "home water," the Ten Mile River of New York state, near the Connecticut border. This is where Tanner learned for the first time "that loving a body of water meant that you took the bad with the good, that you felt it personally when your favorite pools filled with silt from the new housing developments on the banks, and the trout and the bass all disappeared. I fished the Ten Mile day in and day out for years, and I eventually came to understand that for every ten new homes in the riparian zone, there'd be another half inch of silt on the bottom of the river."

We often underestimate the power of caretaking even the smallest patch of ground to inspire awareness and action in ourselves and others. In *The Summer Book*, in which she describes a remote island in the Gulf of Finland visited by an elderly woman and her six-year-old granddaughter, Finnish author Tove Jansson wrote of the grandmother: "Even though she knew there was no need to feel sorry for small islands, which can take care of themselves, she was uneasy whenever there was a dry spell. In the evening she would make some excuse to go down to the marsh pond, where she had hidden a watering can under the alders, and she would scoop up the last dregs of water with a coffee cup. Then she would go around and splash a little water here and there on the plants she liked best, and then hide the can again. Every autumn, she collected wild seeds in a matchbox, and on the last day she would go around and plant them, no one knew where."

Although many residents around Lake Leelanau share Gloria's appreciation of the land, her initial efforts to teach mindfulness in nature practices, as she had in California, did not take hold. Gloria pivoted and began training to be a shoreline consultant with the Lake Leelanau Lake Association. The goal of the association is to help others with shoreline protection and riparian area beautification projects that are compatible with overall landscape health.

Many landowners on the Leelanau Peninsula delight in creating large lawns, lavish homes, and extravagant boathouses. Although Gloria admits that these creations showcase inspiring architecture, they represent a disconnect between the land and its broader sweep of residents. Well-fertilized lawns, for instance, come with runoff that gives rise to unpleasant algal blooms in nearby lakes and ponds. Remedying this kind of unintended consequence is part of nature's conversation too. The verdant lawns also increase the number of ducks and geese feeding on those shores, and along with them comes swimmer's itch. The itch is an allergic skin reaction to certain microscopic parasites that proliferate in shallow waters with lots of snails and waterfowl. When landowners are ignorant of the underlying cause and effect, they perpetuate the itchy problem for their children and grandchildren.

Gloria openly admits that she has gone through an evolution in her own thinking. "There are things I did twenty-five years ago that I would never do today... Humility is an essential part of the process." As Gloria's dog-eared copy of Robin Wall Kimmerer's *Braiding Sweetgrass* suggests, creating a garden is one of the best ways to give back to yourself and to reconnect with the land. Equally rewarding in a slightly different way is tending a wild garden, as the grandmother in Tove Jansson's novel did. Rather than creating a plot where you tend the land expressly for the harvest of fruits and vegetables, reenvision your wild garden as a place where you can cultivate relationships by assisting native flora and fauna. This alternative type of garden can be a source of immense happiness. With all that is going sour around the planet, it is empowering to know that there is one spot where each of us can do something positive, right now.

In urban Melbourne, Australia, Gio Fitzpatrick has set out to rehabilitate the old Elsternwick golf course. The goal is to "re-create the pre-colonial wetlands, right in the heart of an urban community," he says in an article published in the *Guardian* featuring projects that are reclaiming slivers of ground and water lost to urban growth.

"Within a few years, we've gone from zero fish to counts in the hundreds," says another land steward from a project in Queensland. "Eels, turtles, spoonbills, all kinds of birds. There's now even some resident kangaroos." And plants are benefiting too. In Sydney, where a light-rail project was established, a volunteer noted, "Where [we've] gotten rid of the weeds the native species have a chance to breathe and grow now. And we're actually still discovering plants that would have been in the forest before the settlers came."

Fitzpatrick notes that convincing the community to rehabilitate the old golf course has been an uphill battle, but he also says that since work began "visitation has gone through the roof," and "it's now probably one of the most heavily used green spaces in the area. There were a lot of locals that didn't know about the place but now do, and have absolutely fallen in love with it."

A wild garden is also perfect for those who lead a life on the go. For one, many of the things you would weed out of a traditional garden are those you will want to foster in a wild one. Second, in those times when you are housebound or away, you won't have to hire someone to water, weed, or maintain fencing. Tending

wild plots won't require as much hands-on labor as domes-
ticated plots, as they usually handle all the seeding, planting,
pruning, and diversification on their own, though they will do
better if you can pitch in from time to time.

Jenny and I tend our wild garden by picking a few of the
non-native plants, such as houndstongue (*Cynoglossum officinale*)
and madwort (*Asperugo procumbens*), or by removing a few dead
branches from a struggling dogwood shrub on almost every outing. Although we do not rely on the wild gardens around home
for food, we sometimes enjoy a treat of salad greens or herbs
in the form of nettles, dandelions, yarrow, plantain, bitterroot,
and occasional raspberries or strawberries. The fruits of our
labor more often take the form of improved soil quality, greater
plant and animal biodiversity, and more insects, including those
all-important caterpillars mentioned earlier, which help nesting
birds and other animals.

As with your sit spot, make sure your wild garden is as close
as possible so you visit it regularly. It doesn't have to be big;
four- or six-feet square is more than enough to start. If you can't
establish a garden adjoining your house or apartment, cultivate
one in an open lot, a patch of the city greenbelt, or a county park.
Make it your own from a caretaking perspective and expand the
effort as time and energy allow. And make sure you are in accor-
dance with the local regulations. If for some reason there are
rules preventing you from adopting that land, one of your first
acts of stewardship might be to change those rules. Otherwise,
partner with existing land managers, as they are chronically
underfunded and understaffed.

The work of biologist and author Doug Tallamy shows us how
easy and vital it is to champion our wild friends if we want to

help keep cross-species conversations going. In his book *Nature's Best Hope*, Doug urges us to take matters into our own hands and not wait for governments or NGOs to save our imperiled wildlife. One of the best hopes for nature is to cultivate a wild garden wherever you can. Doug calls this creating your own Homegrown National Park. We all have potential to make something more of the land either by cultivating a wild plot of land beside one's home or by nurturing a natural space in some small corner of a public seashore or city park nearby.

Taking care of the wild ones can also mean something as simple as choosing a different walking route when we know that the nighthawks are nesting on one particular hillside. It is tremendously rewarding to see animals responding to our gestures. A past student of mine from the Upper Midwest altered her approach to morning outings on her paddleboard to accommodate a family of common loons. When she started steering around the nesting area and the brood, the birds responded in kind by hanging out in front of her dock. Her "pet loons" caught the attention of neighbors, who voiced their envy. "Why on earth does she get to have the loons hanging out right in front of her place, and no one else does?" The woman communicated her respect of their space and the family of loons responded.

Becoming involved inevitably means that when you come across places where the local flora and fauna are in a fragile state, you have a responsibility to protect them. As friend and professional nature photographer Dan Hartman puts it, when you find these special, sensitive areas of your home ground, "it's your responsibility." You have been given privileged knowledge and you have to guard it carefully. Anyone who fishes knows loose lips not only sink ships, but also put extra people smack-dab in the middle of sensitive places. One ill-conceived social media post could destroy an area in an instant. If you create a waypoint for others to casually and unwittingly visit, you could spell the end of a very good thing.

Above all, explore responsibly as an active participant rather than as a detached observer. Life is richer and more enjoyable when we are plugged in to the network of connections beyond our human-focused existence. Making the places where we live our own is about giving away bits of ourselves in order to gain membership in something bigger. In our most recent phone conversation, Gloria told me that she heard the most unusual sound coming from a crow through an open window. Curious about what the call might mean, Gloria stepped out onto the deck and looked and listened more closely. Below the perched crow, she suddenly heard a yowling, cat-like cry. She saw not one but two bobcats—a courting pair! Gloria's work is paying off and others are starting to see her home as their home too.

CONCLUSION

NATURE IS CONTINUALLY ASKING US to expand ourselves. How many more layers of the onion can you peel back? Did you hear that? What did that kookaburra just say? Can you smell this? Did you catch that pivot of the rabbit's ear? Noticing these things and connecting them with the bigger picture not only allows us to see what we have been missing, but also offers us the chance to find where we fit in the grand scheme. As we let the outside in, there is almost no way we cannot be changed in the process.

When exploring the calls of the wild, we really do have the chance to see the past, present, and future all at once. Meaning is indeed being communicated in ways that extend beyond the 6,909 languages of humankind. We coevolved as speakers and listeners operating on many of the same fundamental wavelengths as our wild contemporaries. Tone, volume, and context convey animal messages in universal ways that we intuitively respond to at both conscious and subconscious levels.

A friend summarized the importance of direct experiences with the more-than-human planet by saying, "You remember where you were when 9/11 happened, when Kennedy was shot, or when the COVID-19 lockdown began." These momentous events create lasting memories. But if you get all your nature fixes through books and films, odds are you won't remember much. Even if the footage is something "never before filmed," you will forget it within the hour. Alternately, if you have a personal encounter, even if quite mundane by network TV standards—like a hawk buzzing your head just before it snatches a meal—you will remember it for days, weeks, and maybe for the rest of your life. Direct personal experiences are the threads that stitch us back into Mother Nature's fold.

Eavesdropping on wild conversations is about deciphering meaning in animal expressions, but it is only the beginning. Rekindling multi-species conversations opens the door to a deeper level of kinship. The more-than-human voices also help us heal in unexpected ways. And once you see the patterns hidden in plain sight, you can never go back. Listening to a bit of bird banter or an amphibian chorus also teaches us to be better listeners for friends and family, and might even lead to a broader acceptance of humans who are different from us too. You may find that skills learned from eavesdropping also apply to unexpected spheres such as business, when relating to coworkers and partners. A successful investment-fund manager once approached me after a lecture to say that the skills that I recommended that he and others use to uncover patterns in nature were exactly the same as those he used to track financial trends unrecognized by others—he had built his career on them.

A great deal of what we get from wild conversations involves letting go. We have to let go of ideas that have been trumpeted for centuries, such as our superiority, or that animals don't have language, emotions, memories, culture, and so on. Animals continually surprise us with what they do and what they know. We need to let go of needing to have all the answers. And we need to do away with the idea that wildness exists only in far-off places—wildness, like home, is where you find it. Let go of the busy mind and open yourself up to what is around you.

Identifying a portal or point of entry into the natural world can help you enter the conversation of nature whenever you wish. Establish a few signposts or reminders in the spaces you frequent. A story from a trip abroad transformed several of my own impressions of home, and it might be helpful to you as well.

Our family made a trip to Marrakech, Morocco, and its labyrinthine medina known as Djemaa El Fna some years before the devastating earthquake of 2023. The smell of cooking kebabs, woodsmoke, animals, and spice sellers' stalls all intermingled. We wandered as tourists do, taken in by the snake charmers, monkey handlers, multihued jars of turmeric, cumin, ginger, paprika, and coriander, the rhythmic drumming of street musicians, exquisite jewelry, and vibrant textiles.

Living next to an international-tourism destination ourselves, we felt it was important to find ways to get beyond surface-level impressions, and so we hired a local guide. Abdul led us into corners of the city we would never have found on our own. Walking past the Koutoubia Mosque and numerous elements dating to the twelfth-century Almoravid dynasty, we entered cloistered passageways deeper in the market. We toured the city's bread ovens and centuries-old architecture, including the richly decorated Medersa Ben Youssef, a beautiful old university that thankfully was left unscathed by the 2023 earthquake.

Abdul emphasized that the layout of the city and its construction were anything but random. The archways that led visitors in and out of the marketplace also acted as thoroughfares between different cultures and customs. Each arch linked distinct ethnic traditions, alternate ways of dressing, cooking, and worshipping. The physical symbolized the invisible. As we walked beyond the back alleys of the market and into the maze of branching streets—some so narrow you could not walk through them with outstretched arms—our route led us to neighborhoods where many residents of Marrakech lived.

These residential spaces were filled with more ornate architecture, exquisite mosaics, tired donkeys attached to carts, children playing, and, quite notably, numerous lavishly carved doors. Many of these doors had smaller doors set within them. After pondering the matter, I asked Abdul about the doors. I treasure his response to this day. "The larger door," Abdul said, "is used only for large community events such as funerals and weddings. The smaller door is for everyday use. The smaller door makes you bow as you enter." Bowing to step through is a physical act of acknowledgment and humility. Each space demands a new kind of awareness. The doors acted like a scaled-down version of the arched gates of the medina.

Following our return from North Africa, I stopped dead in my tracks on a morning walk. Just before the trail opened into Wet Meadow, a series of branches formed a low archway between an aged aspen tree on one side and a dense, squat juniper on the other. I had always subconsciously shifted gears as I entered Wet Meadow. Now, after pondering Abdul's words, I was conscious in a new way. Like the inner doors of Marrakech, the sticks forced me to lower my head and enter the meadow with a renewed sense of awareness and reoriented expectations.

If I manage to get all the way to Wet Meadow without stepping outside my own thoughts, this minimalist doorway of dead branches ensures that I slow down, take notice, and be reverent.

What would be considered an obstacle by many trailblazers has become a meaningful threshold that our family now maintains. If a storm or exceptionally large buffalo brings down some of the archway, like the residents of Marrakech lifting toppled walls and collapsed minarets of their city back into place, we replace the missing wood. The upkeep is not for any structural role it plays, but because of what it stands for.

What will your passageway be? What are you overlooking and not hearing because of the dialogue going on inside your head? To hear the quietest voices and see the subtlest gestures of nature's discourses, we must reside in the now. Establish some of your own doorways as reminders. These things don't have to be big or obvious. They could amount to a uniquely colored stone, a notable tree, or even your actual doorway. Better still, establish several reminders. A particular street intersection could serve as a portal. Stepping around a plant can become a doorway. "Take off the blinders," the portals will say. "Be deliberate about those you meet and how you choose to meet them. Listen to what's being said."

I wish that when I walked out the front door the birds didn't flush or the deer didn't start walking away. It would be great to know the meaning of each and every chirp or howl in the ways that the First Peoples of this place did—the Crow, Shoshone, Kiowa, and Nez Perce—but I try. Each day, it helps to go just a little bit beyond where I left off the day before.

The harder we work, the longer the to-do list, and the more gadgets we gather around us, the more we need wild experiences. Flush away the anthropogenic clutter of modern living—even if only for a minute or two each day—by diving into the cool waters of a listening break. Our evolved affinity for animal voices makes them uniquely suited to helping us push the reset button—they add understanding, calm, clarity, and happiness to our days.

Explore what interests you. Investigate what fills your cup and gives you joy. Cultivate trust with the wild ones and, wherever possible, become a gardener of dialogue and sow seeds for wider conversation. Strive to become the world's leading expert on the place that you call home. Fine-tune a sensitivity to the places that feed you physically, mentally, emotionally, and spiritually and in turn urge you to honor that gift by giving of yourself.

Although Jenny, George, and I reside on the edge of the world's first national park, my treasured sit spot, my reset button, and my wild gardens—my *real* classrooms—rest within twenty-five steps of our door. Open up to seeing nature not as a theoretical construct or newspaper column, and don't just look—step into the tangle of relationships with the whole of your being. And the next time you walk through a chosen portal and hear a bird or squirrel chattering, pause long enough to ask, "Are you talking to me?"

ACKNOWLEDGMENTS

WRITING THIS BOOK has been a wonderful journey and I am humbled by those who have joined me along the way. In particular, I would like to thank my friend Jeff Reed for the use of his mountain cabin as a writing space, for our shared adventures, and for endless dialogues on the languages of humans and nonhumans alike. Thank you to Brad Orsted for our many walks and conversations on nature, life, and writing. My sincerest gratitude goes out to Jon Young for educating us all about animal language, for his stories, for our philosophical exchanges, and, of course, for his friendship. Thank you also to Marc Bekoff and Joe Hutto for your inspiring work that addresses many things science struggles to define. Thank you to Casey Anderson, Shauna Baron, Rick Bedsworth, Colette Daigle-Berg, Judith DeVito, Chuck Ernst, Guy Galante, Gloria Garrett, Erick Greene, Jack Horner, Matthias Loretto, John Marzluff, Fred Newman, Kate and Adam Rice, D. Andrew Saunders, Lynelle Schuerr, Katie Sieving, Con Slobodchikoff, Lauri Travis, Kent Weber, and Carol Wing, who graciously allowed me to tell their stories. I would also like to dedicate the book to the memories of Floyd DeWitt and Kevin Reeve—your lessons will stay with me until the end of my days.

This book would not have been possible without the influence of many others, including the late Dick Wallace, Carl Stallknecht, Bill Burnell, Kenny Osmond, Frank Dunham, Ken and Harriett Cooper, Jim Dougherty, and Jim Langell. Thank you also, in roughly geographic order, to Scott Cheney, William Burnell, Betty Osmond, Jim Montanaro, Rick Nicolini, Jennifer Scanlon, Joe Dembeck and Natalie Jones, Casey Burns, Rainer Brocke, Robin Wall Kimmerer, Larry VanDruff, Jason Wilson and Debbie Lineweaver, Clyde Kessler, Tim and Sylvia Copeland, Rachel Gray, Dean Stauffer, Chris Swank, Mike Vaughan, Gene Ball, Isabel Behnke, Savannah Rose, Len Carolan, Bob Crabtree, LaRita Desimpel, Shane Doyle, Katy Duffy, Dawson Dunning, Gary Ferguson, Jim Garry, Steve Gehman and Betsy Robinson, Kerry Gunther, James Halfpenny, Bill Hamblin, Dan and Cindy Hartman, Jeff Hogan, Bill and Peggy Hoppe, Bob Landis, MacNeil Lyons, Emil McCain, Casey McFarland, Rick McIntyre, Gerry Ohrstrom, Dale and Elva Paulson, Brian Kahn, Kirsty and Alan Peake, Eric Reinertson, Jennifer Sheldon, Doug Smith, Dan Stahler, Bob Weselmann, Nathan Varley, Robert Glen and Sue Stolberger, Kim Baron, Bob and Birgit Bateman, Michan Biesbroek, Nicholas Reti and Tracy Schwartz, Julie Zickefoose, Wendy Davis, Karl Gegenfurtner, Gary Gilbert, David George Haskell, Bernd Heinrich, Janet Kessler, Bernie Krause, Phillip Marino, Jose M. Amoedo, Lee Burton, Claire Dunn, Kristi Dranginis, Dan Gardoqui, Lonner Holden, Josh Lane, Erika Larsen, Mandylee Crampton, Nathan Pieplow, Nate Summers, and the library staff at the Yellowstone Heritage and Resource Center.

To Jane Billinghurst, my editor, I cannot say enough. Had you not asked the question "Have you ever thought of putting this material into a book?" this volume would still be buried in overstuffed folders and piles of course notes. Your thoughtful guidance has elevated this book severalfold. Thank you to Bill Bridgeland and Robin Rauch, my dear wife, Jenny Golding, Bianca Klein, Joanna Lambert, Katy Newman, and Mike Twist—your comments added immense clarity and integrity to the manuscript. The Montana Arts Council also provided support for the research and development of this book.

Most of all, I would like to thank my lifelong teachers: Raven, Muskrat, Momma Deer, the wild turkeys, Beaver, bull elk #6 and #10, Bernese the ruffed grouse, and wolves 42F, 21M, and 302M. Thank you also to Bison and Otter, Rocky Mountain parnassian and Gillette's "Yellowstone" checkerspot butterflies. I am indebted to Easton and Weston the song sparrows, Scarface and 264 the grizzly bears, Coyote and Magpie, Bent Toe the sandhill crane, Red and Gray Fox, the bluegills, sunfish, perch, and bass of my youth, Carp and Bowfin, Bull Snake and Bullfrog, *Hexagenia*, Canada Goose, *Dytiscus*, Wood Duck, *Lethocerus*, and *Pteronarcys*. I hope I have honored your gifts and given you a voice when so many have failed to hear; hopefully, after reading this, more humans will find their way into your master courses.

And to you, the land, with my eyes nearly brimming over—Big Bay Swamp and Oneida Lake, my home waters. Thank you also to the upper Genesee River and Tioughnioga Wildlife Management Area, where the deer taught me much, to Tom's Creek and the Brown Farm, Clinch Mountain Wildlife Management Area, Eagle Creek, the sprawling glacial floor of Lamar Valley, Pebble and Slough Creeks, and Specimen Ridge. You are always with me. Finally, I thank the countless generations who stewarded my friends, teachers, and the land for the thousands of years before I joined this Earth—may your wisdom help carry us to a more inclusive, compassionate, and constructive future for all.

APPENDIX: PRACTICES FOR EVERY DAY

This appendix summarizes the specific suggestions given in this book for entering the world of creature-speak. You can advance your practice by accessing further resources, including audio and video recordings, at GeorgeBumann.com/Eavesdropping.

CHAPTER 1—VOICES ALL AROUND

- Put down whatever you are doing right now, step outside, and listen. Listen to everything, including things you usually block out, including the sounds made by other people (automobiles and conversations), the Earth itself (the ocean surf or thunder), and, of course, the animals.

CHAPTER 2—ENTERING CREATURE-SPEAK

- Identify the spaces where you spend most of your time. What parts of these spaces are best suited to observing wild conversations? Is your office desk next to the window? Can you hear what's going on beyond those open cargo doors?

- Make a list of common tasks that you regularly perform and then ask yourself, "How many of these can I do outdoors?" Can you take phone calls on the patio? Can you eat lunch along the parkway? Actively and passively draw your attention to what is normal in your area.
- Concentrate on patterns and tendencies, rather than on the identification of species.

CHAPTER 3—HEARING A PIN DROP

- Become a dedicated student of nature. Peer further into the details of life by reviving your childlike wonder. Select a natural object to draw. Document what you find using your creative side to both draw your attention to unseen things and record them for future reference.
- Engage the faculties you have—better. Reignite your senses in a focused way. Practice using your peripheral vision to its fullest potential. Did you catch that fly buzzing around out of the corner of your eye? Close your eyes and point to the location of a sound, then take a look to see how close you are.

CHAPTER 4—THE SIGNAL IN THE NOISE

- Start noticing basic animal patterns across time and space. Where exactly do the flocks of birds hang out or those swarms of insects emerge? Turn your thinking toward what constitutes "normal." Are animals gathering in unusual numbers? What are they doing? When do they arrive and when do they leave? Are the things you do influencing this pattern?
- Put a mark on the calendar or, if you post observations on social media, revisit your archived updates from months and years past to compare the present with the past. Are the ducks

in eclipse plumage, and are the garden spiders spinning webs in the same places?

CHAPTER 5—MAKING SENSE OF MINUTIAE

- Look for the little things close at hand, starting with insects, spiders, or other invertebrates.
- Find ways that these little things connect to the big picture through the choice of food or who feeds on them. How does this change across the seasons? Can you identify a butterfly by the way it flies, or the location of a mammal scat by where a butterfly alights?
- Conduct a few experiments with your gardening or lawn care practices, perhaps by setting aside a part of your lawn to be unmown and seeing how it differs from mown sections. How does the pollinator population compare from year to year or with national or global trends?

CHAPTER 6—ANIMAL SENSES

- Read up on the super-senses of your pets and most common wildlife, then take what you learn outside. Follow an animal around and do your very best to notice what they are noticing. What grabs their attention? What sets them at ease? And how many of these things are within your own abilities to detect?
- Use this as a stepping-off point to realizing how their senses can inform your own.

CHAPTER 7—OUR SENSES

- Discover how much your senses are really capable of detecting. Set aside your perceived limitations and elevate your lesser senses to new heights. Try sniffing your way through a pet store, a zoo, or someone's home. Listen for sounds in the highest or lowest frequency ranges by taking a hearing test on YouTube (www.youtube.com/watch?v=-E1SDl9vLo8) or strive to describe a new flavor in exquisite detail, then see how many other places it shows up.

CHAPTER 8—LEARNING BY DOING

- Improve your sensitivity to wild beings by imitating them. Tap out a woodpecker drum solo with your finger or become a stalking cat to recalibrate your neural pathways. Sketch a visual representation of the sound, apply phonetics to write out a memorable phrase from a bird's tune, or jot the notes down on a scale. Involve as much of your mind, body, and abilities as possible.

- Like Fred Newman, set aside a few minutes each day for a listening break to recalibrate your senses, relax, and find even more of what you've been missing.

CHAPTER 9—YOU'RE BEING WATCHED

- Explore how wild species might recognize you, friends, family, and neighbors as unique individuals.
- Discover how you, specifically, are being perceived. What is your current rapport with these others? When birds are waiting for you to fill the feeder, do they take flight, stay put, or approach?
- Consider what kind of relationship you want going forward.

CHAPTER 10—MAKING WAVES

- Identify a good place to sit and observe, with a low level of human traffic. Take up your post for an hour or more with no agenda beyond observing animals great and small.

- Wait for the disruption of your own arrival to subside, then pay attention to waves of disruption created by others.
- Watch for unified patterns of movement in multiple species.

CHAPTER 11—SCARIEST OF THE SCARY

- Observe how animals react to kind, benign humans versus unkind ones.
- Notice how close people are allowed to get before the wild observers get uneasy. How far up into the trees do they go to get away? What distance do they run before stopping and looking back?
- Use yourself as the test subject and try acting in a benevolent or exceedingly kind and deferential way to all the creatures you meet. Give area nonhumans a wide berth. Be hypersensitive to their every move, withdraw immediately whenever they show signs of stress, talk to them gently, lovingly even. Do they change their tune?

CHAPTER 12—DECODING THE LANGUAGE OF ALARM

- Ask yourself what the ultimate and proximate causes might be of animal behaviors. What are the limiting factors in your immediate area? Is it water, food, or space? How do these factors influence what local species do and don't do?
- Listen to the howls of local coyotes, wolves, or domestic dogs. Do they vary in length, inflection, or intensity? Do they include a lot of barking? If you can spot the barker (or an alarmist of another species), look in the direction they are looking. Is the animal also using the shadowing approach to get a better fix on where the trouble might be?

CHAPTER 13—SPEAKING CHICKADEE-ESE

- Try counting *dees* to gain insight into the lives of local chickadees. Know that birds and other animals will be eavesdropping and giving their own calls.
- Listen for sustained calling by multiple gossips in a single location and go see what all the fuss is about!
- Make note if you hear high-pitched *seet* calls and/or see any birdsicles.

CHAPTER 14—VARIATIONS ON A THEME

- Disentangle finer layers of context and meaning by studying one common call that you hear from the most common wildlife species. How does it modulate? What's going on then?
- Take notes so you can refer to them later; imitate the calls, or record them using the video or audio functions on your phone. If things get complicated, try creating a vocabulary list in your journal or on your cell phone.

CHAPTER 15—HOWLING MYSTERIES

- Tackle the mysteries of life by hitting the books or, easier still, the internet search bar. Have others seen what you've observed? Has it been studied? Although the internet can be full of erroneous information and false leads, the web can refer you to reliable sources such as films, newspaper articles, books, and research publications. For the latter, you may find tools such as Connected Papers (www.connectedpapers.com) helpful to expand your searches to related but hard-to-locate material.

CHAPTER 16—1 + 1 = 1,000

- Advance your understanding of animal language in one of the quickest ways possible—a group bird sit. Organize as many other participants as you can and debrief the sit as thoroughly as you have time for. Do this on a regular basis for the best results.

- For guidelines on how to structure and debrief a group bird-language sit, consider purchasing the *Bird Language Basics* video by Jon Young (villagevideo.org/birdlanguage).

CHAPTER 17—THE HONEYCOMB LANDSCAPE

- Strive to spend thirty to sixty minutes, or more, in a good sit spot.

- See if you can identify a patch of silence, even if it is the one you created when walking to your spot.

- Can you delineate the edges of that silence? Look for woodpeckers or other sentinels for clues. How big is it? How long does it last, and does it move? Observe how other creatures respond to the silent void.

CHAPTER 18—WILDSENSE 2.0

- Cultivate trust and enjoy closer proximity to wild creatures by demonstrating acts of recognition and respect. Practice your own kind of deference by giving animals room to carry out essential tasks. Pause when you encounter them, back up if necessary, avoid eye contact, and if you speak, do so in a soft, comforting tone.

- Learn the species-level and individual-specific cues of discomfort and respond to them. Adjust your own behavior until you see the animal exhibit relaxing markers and note how they respond.

- Play by the rules, as some localities, such as national parks, require setback distances from animals.

CHAPTER 19—BECOMING INVISIBLE

- Reorganize your subconscious messaging with a meditative practice, such as the bridge to silence. Quiet your thinking mind by letting sensations of outward things, such as leaves in the wind or the details of a beautiful flower, completely occupy your attention. Take note of how animals react to you in those relaxed states compared with other times.
- Make a recurring date with yourself to enjoy that state of elevated calm and place it on the calendar; defend that date as you would the most important appointment, such as a long-awaited vacation or a trip to the doctor's office. And when the stress of life takes hold, go to your quiet place, if only in your mind.

CHAPTER 20—THE POWER OF ONE

- Identify animals with unique marks such as an oddly colored feather or shell, a limp, or a nick in its ear. Give that individual a name or a handle to help remember them. Observe how widely they travel and who they hang out with, along with their unique behaviors, preferences, and traits.
- Focus on what's best for your newfound friends, even if that means avoiding interaction or close physical proximity to them. Building a relationship with someone of another kind is a privilege and it comes with substantial responsibilities.

If it ever looks like your time together will do them harm in any way, give them their space and observe from a greater distance.

CHAPTER 21—MAKING IT YOUR OWN

- Create personal names for local landmarks to increase your familiarity with the place you call home. Let the names occur to you through personal experiences that involve close friends, family, and especially children; invoke humor to make them memorable, too!
- Create a wild garden or a Homegrown National Park. Learn the most common plants and whether they are native or non-native, harmful or helpful, to local ecosystems, and steward that plot of ground to benefit area wildlife.

CONCLUSION

- Identify portals or points of interest along daily routes as reminders to be more aware, receptive, and in the present moment. This could be an actual doorway, a garden arbor, a street intersection, or a unique building, rock, or tree.

NOTES

Topics are listed in the order in which they appear in each chapter; references within topics follow the order in which individual concepts are discussed.

CHAPTER 1: VOICES ALL AROUND

Anthropause: C. Rutz, M. C. Loretto, A. E. Bates, et al., "COVID-19 Lockdown Allows Researchers to Quantify the Effects of Human Activity on Wildlife," *Nature Ecology & Evolution* 4 (2020): 1156–1159, doi.org/10.1038/s41559-020-1237-z.

Wild boars in Rome: D. Bressan, "Animals Roam Freely in Italian Cities With Humans in Lockdown," *Forbes*, March 20, 2020, www.forbes.com/sites/davidbressan/2020/03/20/animals-roam-freely-in-italian-cities-with-humans-in-lockdown/.

Sparrows in San Francisco singing different tunes: "Scientists Find the Quiet of Pandemic Shutdowns Has Made Birds Change Their Tunes," *All Things Considered*, aired September 24, 2020, on NPR, www.npr.org/2020/09/24/916625364/scientists-find-the-quiet-of-pandemic-shutdowns-has-made-birds-change-their-tune.

Leopards on outskirts of Bengaluru: A. K. Behera, P. R. Kumar, M. M. Priya, et al., "The Impacts of COVID-19 Lockdown on Wildlife in Deccan Plateau, India," *Science of the Total Environment* 822 (May 20, 2022): 153268, doi.org/10.1016/j.scitotenv.2022.153268.

CHAPTER 2: ENTERING CREATURE-SPEAK

Human turkey calls versus actual turkey calls: D. Collins, "Callers Sound Better Than the Turkeys," *Washington Post*, March 13, 1984, www.

washingtonpost.com/archive/sports/1984/03/13/callers-sound-
better-than-the-turkeys/15f6602d-ec4e-4fe7-b5a8-626e78bdd0b9/.
David Brooks: D. Brooks, *The Social Animal: The Hidden Sources of Love, Character, and Achievement* (New York: Random House, 2011).

CHAPTER 3: HEARING A PIN DROP

Oldest animal art: Adam Brumm, Adhi Agus Oktaviana, Basran Burhan, et
al., "Oldest Cave Art Found in Sulawesi," *Science Advances* 7, no. 3 (2021),
doi.org/10.1126/sciadv.abd4648.
Blind observers during World War II: Britain: War Office–Signals Experimental Establishment, *Experiments With Blind Listeners*, report no. 303
(1925).
Hearing of blind versus sighted people: Abdulrahman Ali Al Saif, Abdulrahman Khalid Al Abdulaali, Kamal-Eldin Ahmed Abou-Elhamd,
and Sayed Ibrahim Ali, "Do Blind People Have Better Hearing
Levels Than Normal Population?" *International Journal of Advanced
Research* 5 (January 2017): 3004–3009, www.academia.edu/76497812/
Do_blind_people_have_better_hearing_levels_than_normal_
population. N. Lessard, M. Paré, F. Lepore, and M. Lasonde, "Early-Blind
Human Subjects Localize Sound Sources Better Than Sighted Subjects,"
Nature 395 (1998): 278–280, doi.org/10.1038/26228.

CHAPTER 4: THE SIGNAL IN THE NOISE

Timing of bird and amphibian choruses: T. J. Brown and P. Handford, "Why
Birds Sing at Dawn: The Role of Consistent Song Transmission," *Ibis* 145
(2003): 120–129, doi.org/10.1046/j.1474-919X.2003.00130.x. E. D. Bel-
lis, "The Influence of Humidity on Wood Frog Activity," *The American
Midland Naturalist* 68, no. 1 (1962): 139–148, doi.org/10.2307/2422640.
Rodrigo Lingnau and Rogério P. Bastos, "Vocalizations of the Brazilian
Torrent Frog *Hylodes heyeri* (Anura: Hylodidae): Repertoire and Influence of Air Temperature on Advertisement Call Variation," *Journal
of Natural History* 41 (2007): 17–20, 1227–1235, doi.org/10.1080/
00222930701395626.
Molting waterfowl: J. K. Ringelman, "13.4.4. Habitat Management for Molting Waterfowl" in US Fish and Wildlife Service, *Waterfowl Management
Handbook* (Lincoln: University of Nebraska, 1990), digitalcommons.unl.
edu/icwdmwfm/24.

CHAPTER 5: MAKING SENSE OF MINUTIAE

Spider silk: L. Brunetta and C. L. Craig, *Spider Silk: Evolution and 400 Million Years of Spinning, Waiting, Snagging, and Mating* (New Haven, CT: Yale University Press, 2010).

Spiders recolonizing Krakatoa: David Quammen, *The Song of the Dodo: Island Biogeography in an Age of Extinction* (New York: Scribner, 1996).

Reduced nectar sources in urban and suburban environments: Xerxes Society, Bee City USA, "No Mow May, Low Mow Spring" (2023), beecityusa.org/no-mow-may/.

Butterflies acquiring salts: J. A. Scott, *The Butterflies of North America: A Natural History and Field Guide* (Redwood City, CA: Stanford University Press, 1992).

CHAPTER 6: ANIMAL SENSES

Olfaction in humans: B. Handwerk, "In Some Ways, Your Sense of Smell Is Actually Better Than a Dog's," *Smithsonian Magazine*, May 22, 2017, www.smithsonianmag.com/science-nature/you-actually-smell-better-dog-180963391/. J. P. McGann, "Poor Human Olfaction Is a 19th-Century Myth," *Science* 356, no. 6338 (2017), doi.org/10.1126/science.aam7263.

Sharks detecting weak electrical fields: E. Yong, *An Immense World: How Animal Senses Reveal the Hidden Realms Around Us* (New York: Random House, 2022).

Seals following fish by disturbances in water: K. Knight, "Harbour Seal Whiskers Detect Fish Trails 35 s Later," *Journal of Experimental Biology* 213, no. 13, doi.org/10.1242/jeb.047258.

Shedding of skin cells: Imperial College London, "New Insights Into Skin Cells Could Explain Why Our Skin Doesn't Leak," ScienceDaily, November 29, 2016, www.sciencedaily.com/releases/2016/11/161129114910.htm.

What cadaver dogs smell: J. Neimark and Sapiens, "The Dogs That Sniff Out 5,000-Year-Old Bones: Archaeologists Are Using Canine Assistants to Uncover Ancient Remains," *Atlantic,* July 5, 2020.

Dog olfaction: D. B. Walker, J. C. Walker, P. J. Cavnar, et al., "Naturalistic Quantification of Canine Olfactory Sensitivity," *Applied Animal Behaviour Science* 97, nos. 2–4 (2006): 241–254. A. Greenberg, "6 Stinking Cool Facts About Dog Noses: Dogs Can Sniff Out Disease and Analyze New Odors Even as They Exhale. But How?" *NOVA* online, June

10, 2022, www.pbs.org/wgbh/nova/article/dog-smell-olfaction-nose/.

J. Levine, "The Education of a Bomb Dog," *Smithsonian Magazine* online, July 2013, www.smithsonianmag.com/innovation/the-education-of-a-bomb-dog-4945104/. C. Fischer-Tenhagen, L. Wetterholm, B. Tenhagen, and W. Heuwieser, "Training Dogs on a Scent Platform for Oestrus Detection in Cows," *Applied Animal Behaviour Science* 131, nos. 1–2 (2011): 63–70, doi.org/10.1016/j.applanim.2011.01.006.

Elephants receive long distance communication through feet: "How Elephants Listen... With Their Feet," *Deep Look*, season 5, episode 13 (2018), www.pbs.org/video/how-elephants-listen-with-their-feet-40owt4.

Singing mice: R. Dunn, "The Mystery of the Singing Mice: A Scientist Has Discovered That High-Pitched Sounds Made by the Small Rodents Could Actually Be Melodious Songs," *Smithsonian Magazine*, May 2011.

Dolphins identifying objects and materials at a distance: C. Wei, M. Hoffmann-Kuhnt, W. W. L. Au, et al., "Possible Limitations of Dolphin Echolocation: A Simulation Study Based on a Cross-Modal Matching Experiment," *Scientific Reports* 11, no. 6689 (2021), doi.org/10.1038/s41598-021-85063-2.

Eel use of electricity: G. K. H. Zupanc and T. H. Bullock, "From Electrogenesis to Electroreception and Overview," in T. H. Bullock, et al. (eds.), *Electroreception* (New York: Springer, 2005): 5–46.

CHAPTER 7: OUR SENSES

Better peripheral vision in the deaf: Daphne Bavelier, Andrea Tomann, Chloe Hutton, et al., "Visual Attention to the Periphery Is Enhanced in Congenitally Deaf Individuals," *Journal of Neuroscience* 20, no. 17 (2000): RC93.

Elevated sound memory in the blind: B. Röder, F. Rösler, and H. J. Neville, "Auditory Memory in Congenitally Blind Adults: A Behavioral-Electrophysiological Investigation," *Cognitive Brain Research* 11, no. 2 (2001): 289–303, doi.org/10.1016/S0926-6410(01)00002-7.

Color-blind vision and camouflage: M. J. Morgan, A. Adam, and J. D. Mollon, "Dichromats Detect Colour-Camouflaged Objects That Are Not Detected by Trichromats," *Proceedings: Biological Sciences* 248, no. 1323 (1992): 291–295, doi.org/10.1098/rspb.1992.0074.

E. O. Wilson's blindness and career path: E. O. Wilson, *Naturalist* (Washington, D.C.: Island Press, 1994).

Jack Horner: L. Attebery, "Success Stories: Jack Horner, Paleontologist,"
Yale Center for Dyslexia and Creativity, 2017, dyslexia.yale.edu/story/
jack-horner.
Temple Grandin TED Talk: T. Grandin, "The World Needs All Kinds of
Minds," TED Talks, uploaded February 2010, www.ted.com/talks/
temple_grandin_the_world_needs_all_kinds_of_minds.
Creativity inspired by nature: D. Alexander, "Biomimicry: 9 Engineering
Innovations Inspired by Nature's Design," *Interesting Engineering,* April
19, 2023, interestingengineering.com/science/biomimicry-9-ways-
engineers-have-been-inspired-by-nature. N. A. Siddiqui, W. Asrar,
and E. Sulaeman, "Literature Review: Biomimetic and Conventional
Aircraft Wing Tips," *International Journal of Aviation, Aeronautics, and
Aerospace* 4, no. 2 (2017), commons.erau.edu/cgi/viewcontent.cgi?
article=1172&context=ijaaa. D. Thomas, "Primavera and the Seasons:
How Nature Influences Music," *Interlude,* August 3, 2020, interlude.hk/
primavera-and-the-seasons-how-nature-influences-music.
Information processing by the conscious and unconscious mind: E. Kwong,
"Understanding Unconscious Bias," *Short Wave,* aired July 15, 2020, on
NPR.
Scent of birds: D. J. Whittaker, *The Secret Perfume of Birds: Uncovering the Sci-
ence of Avian Scent* (Baltimore: Johns Hopkins University Press, 2022).
Rhinos killed by elephants: Rob Slotow, Dave Balfour, and Owen Howi-
son, "Killing of Black and White Rhinoceroses by African Elephants in
Hluhluwe-Umfolozi Park, South Africa," *Pachyderm* (2001): 14–20.
Fifty-four human senses: M. J. Cohen, *Reconnecting With Nature: Finding
Wellness Through Restoring Your Bond With the Earth* (Lakeville, MN: Eco-
press, 1997).

CHAPTER 8: LEARNING BY DOING
Wolf howling in Algonquin Park: The Friends of Algonquin Park, "Public
Wolf Howls," 2023, www.algonquinpark.on.ca/visit/programs/wolf-
howls.php.
Vocal learning in woodpeckers: Eric R. Schuppe, Lindsey Cantin, Mukta
Chakraborty, et al., "Forebrain Nuclei Linked to Woodpecker Territorial
Drum Displays Mirror Those That Enable Vocal Learning in Songbirds,"
PLoS Biology 20, no. 9 (2022), doi.org/10.1371/journal.pbio.3001751.
Musical scales used by birds: Myrna Oliver, "Luis F. Baptista; Ornithologist
Mastered the Dialects of Birds," *Los Angeles Times,* June 15, 2000, www.
latimes.com/archives/la-xpm-2000-jun-15-me-41262-story.html.

Sound and the brain: University of Queensland Australia, "The Limbic System," qbi.uq.edu.au/brain/brain-anatomy/limbic-system. N. J. Kreutzer, "The Limbic System and Its Role in Affective Response to Music," *Update: Applications of Research in Music Education* 10, no. 1 (1991): 19–24, doi.org/10.1177/875512339101000105.

Gwich'in imitating ravens to hunt bears: R. K. Nelson, *Hunters of the Northern Forest: Designs for Survival Among the Alaskan Kutchin*, 2nd ed. (Chicago: University of Chicago Press, 1986).

Pishing birds: P. Dunne, *The Art of Pishing: A How-To Guide* (Mechanicsburg, PA: Stackpole Books, 2006).

The chameleon effect: T. L. Chartrand and J. A. Bargh, "The Chameleon Effect: The Perception–Behavior Link and Social Interaction," *Journal of Personality and Social Psychology* 76, no. 6 (1999): 893–910, doi. org/10.1037/0022-3514.76.6.893.

CHAPTER 9: YOU'RE BEING WATCHED

Animals remembering others of their species: K. M. Kendrick, A. P. da Costa, A. E. Leigh, et al., "Sheep Don't Forget a Face," *Nature* 414, no. 6860 (2001): 165–166, doi.org/10.1038/35102669. B. J. Pitcher, R. G. Harcourt, and I. Charrier, "The Memory Remains: Long-Term Vocal Recognition in Australian Sea Lions," *Animal Cognition* 13 (2010): 771–776, doi.org/10.1007/s10071-010-0322-0. S. J. Insley, "Long-Term Vocal Recognition in the Northern Fur Seal," *Nature* 406, no. 6794 (2000): 404–405, doi.org/10.1038/35019064. Christine Dell'Amore, "Dolphins Have Longest Memories in Animal Kingdom," *National Geographic News*, August 6, 2013, www.nationalgeographic.com/animals/article/130806-dolphins-memories-animals-science-longest.

"Lower" organisms recognizing one another: A. Avarguès-Weber, "Face Recognition: Lessons From a Wasp," *Current Biology* 22, no. 3 (2012): R91–R93, doi.org/10.1016/j.cub.2011.12.040. D. Baracchi, I. Petrocelli, L. Chittka, et al., "Speed and Accuracy in Nest-Mate Recognition: A Hover Wasp Prioritizes Face Recognition Over Colony Odour Cues to Minimize Intrusion by Outsiders," *Proceedings: Biological Sciences* 282, no. 1802 (2015): 20142750, doi.org/10.1098/rspb.2014.2750.

"Lower" organisms recognizing individual humans: C. Newport, G. Wallis, Y. Reshitnyk, and U. E. Siebeck, "Discrimination of Human Faces by Archerfish (*Toxotes chatareus*)," *Scientific Reports* 6, no. 27523 (2016), doi. org/10.1038/srep27523. Adrian G. Dyer, Christa Neumeyer, and Lars Chittka, "Honeybee (*Apis mellifera*) Vision Can Discriminate Between and Recognise Images of Human Faces," *Journal of Experimental Biology* 208, no. 24 (2005): 4709–4714, doi.org/10.1242/jeb.01929. Sy Montgomery, *Soul of an Octopus*, reprint edition (New York: Atria Books, 2016).

American crows remembering individual people: J. M. Marzluff, J. Walls, H. N. Cornell, et al., "Lasting Recognition of Threatening People by Wild American Crows," *Animal Behaviour* 79 (2010): 699–707, doi. org/10.1016/j.anbehav.2009.12.022. H. N. Cornell, J. M. Marzluff, and S. Pecoraro, "Social Learning Spreads Knowledge About Dangerous Humans Among American Crows," *Proceedings of the Royal Society B* 279, no. 1728 (2012): 499–508, doi.org/10.1098/rspb.2011.0957.

US and Russian militaries using marine mammals: *Frontline*, "Navy Dolphins—a Historical Chronology: A Whale of a Business," PBS online, www.pbs.org/wgbh/pages/frontline/shows/whales/etc/navycron. html.

Elephants identifying humans: L. A. Bates, K. N. Sayialel, N. W. Njiraini, et al., "Elephants Classify Human Ethnic Groups by Odor and Garment Color," *Current Biology* 17, no. 22 (2007): 1938–1942, doi.org/10.1016/j. cub.2007.09.060.

CHAPTER 10: MAKING WAVES

Battle of Chancellorsville: Michael Falco, "Chancellorsville May 1863," Civil War 150 Pinhole Project (2015), civilwar150pinholeproject.com/ the-battle-of-chancellorsville.

Hunting with Ishi: Saxon T. Pope, "Hunting With Ishi—the Last Yana Indian," *Journal of California Anthropology* 1, no. 2 (1974), escholarship. org/uc/item/02r6j5s0.

CHAPTER 11: SCARIEST OF THE SCARY

Lewis and Clark account of killing a wolf: Gary E. Moulton, ed., *The Journals of the Lewis and Clark Expedition*, volume 4 of 13 (Lincoln: University of Nebraska Press, 1983–2001): 111–113.

Wild boar crossing a river to avoid hunting: Peter Wohlleben, *The Inner Life of Animals: Love, Grief, and Compassion—Surprising Observations of a Hidden World* (Vancouver: Greystone Books, 2017).

Whales forming defensive positions against humans: Thomas Beale, *The Natural History of the Sperm Whale* (London: Effingham Wilson, Royal Exchange, 1839).

Grizzly bear response to humans in Yellowstone: Tyler H. Coleman, "Grizzly Bear and Human Interaction in Yellowstone National Park, Bear Management Areas," PhD dissertation (2012), Montana State University. T. A. Graves, "Spatial and Temporal Response of Grizzly Bears to Recreational Use on Trails," master's thesis (2002), University of Montana.

The "landscape of fear": J. W. Laundre, L. Hernández, W. J. Ripple, "The Landscape of Fear: Ecological Implications of Being Afraid," *Open Ecology Journal* 3, no. 3 (2010): 1–7, doi.org/10.2174/1874213001003030001.

Predators fearing humans: J. A. Smith, J. P. Suraci, M. Clinchy, et al., "Fear of the Human 'Super Predator' Reduces Feeding Time in Large Carnivores," *Proceedings of the Royal Society B* 284, no. 1875 (2017): 20170433, doi.org/10.1098/rspb.2017.0433.

Sound recordings of human voices frighten European badgers: Kate Lunau, "Badgers Are More Scared of the BBC Than Bears," Motherboard: Tech by Vice, July 25, 2016, www.vice.com/en/article/kb7gvv/european-badgers-fear-response-oxford-study-bbc.

Sound recordings of dogs' voices displace raccoons: J. P. Suraci, M. Clinchy, L. M. Dill, et al., "Fear of Large Carnivores Causes a Trophic Cascade," *Nature Communications* 7, no. 10698 (2016), doi.org/10.1038/ncomms10698.

Phantom highways and gas fields impact wildlife: H. E. Ware, C. J. McClure, J. D. Carlisle, J. R. Barber, "A Phantom Road Experiment Reveals Traffic Noise Is an Invisible Source of Habitat Degradation," *PNAS* 112, no. 39 (2015): 12105–12109, doi.org/10.1073/pnas.1504710112. Elizeth Cinto-Mejia, Christopher J. W. McClure, Jesse R. Barber, "Large-Scale Manipulation of the Acoustic Environment Can Alter the Abundance of Breeding Birds: Evidence From a Phantom Natural Gas Field," *Journal of Applied Ecology* 56, no. 8 (2019): 2091–2101, doi.org/10.1111/1365-2664.13449.

Human voices played in Santa Cruz Mountains affect mountain lions to mice: J. P. Suraci, M. Clinchy, L. Y. Zanette, and C. C. Wilmers, "Fear of Humans as Apex Predators Has Landscape-Scale Impacts From Mountain Lions to Mice," *Ecological Letters* 22, no. 10 (2019): 1578–1586, doi.org/10.1111/ele.13344.

Two-minute warning: Jon Young, *What the Robin Knows: How Birds Reveal the Secrets of the Natural World* (Boston: Mariner Books, 2013).

CHAPTER 12: DECODING THE LANGUAGE OF ALARM

Squirrels directing alarms at predators: S. M. Digweed and D. Rendall, "Predator-Associated Vocalizations in North American Red Squirrels (*Tamiasciurus hudsonicus*): To Whom Are Alarm Calls Addressed and How Do They Function?" *Ethology* 115, no. 12 (2009): 1190–1199, doi. org/10.1111/j.1439-0310.2009.01709.x.

Nikolaas Tinbergen approach to animal behavior: N. Tinbergen, "On Aims and Methods of Ethology," *Ethology* 20, no. 4 (1963): 410–433, doi. org/10.1111/j.1439-0310.1963.tb01161.x.

Songbirds and winter cold: A. Lewden, M. Petit, M. Milbergue, et al., "Evidence of Facultative Daytime Hypothermia in a Small Passerine Wintering at Northern Latitudes," *Ibis* 156 (2014): 321–329, doi. org/10.1111/ibi.12142. S. M. Sharbaugh, "Physiological Adaptations for Overwintering by the Black-Capped Chickadee (*Parus atricapillus*) in Interior Alaska (64 Degrees North Latitude)," PhD dissertation (1997), University of Alaska Fairbanks.

Energy demands of bird activity: A. G. Horn, M. L. Leonard, and D. M. Weary, "Oxygen Consumption During Crowing by Roosters: Talk Is Cheap," *Animal Behaviour* 50, no. 5 (1995): 1171–1175, doi.org/10.1016/0003-3472(95)80033-6. K. Oberweger and F. Goller, "The Metabolic Cost of Birdsong Production," *Journal of Experimental Biology* 204, no.19 (2001): 3379–3388, doi.org/10.1242/jeb.204.19.3379.

Increased surveillance by multi-species bird flocks: D. R. Farine, L. M. Aplin, B. C. Sheldon, and W. Hoppitt, "Interspecific Social Networks Promote Information Transmission in Wild Songbirds," *Proceedings of the Royal Society B: Biological Sciences* 282, no. 1803 (2015): 20142804, doi.org/10.1098/rspb.2014.2804. J. Bai, T. M. Freeberg, J. R. Lucas, and K. E. Sieving, "A Community Context for Aggression? Multi-species Audience Effects on Territorial Aggression in Two Species of Paridae," *Ecology and Evolution* 11, no. 10 (2021): 5305–5319, doi.org/10.1002/ece3.7421.

Darwin on barking of wild dogs: Charles Darwin, *The Expression of the Emotions in Man and Animals* (London: Penguin Classics, 2009). Eloïse Déaux, "Dingoes Do Bark: Why Most Dingo Facts You Think You Know Are Wrong," The Conversation, December 27, 2016, theconversation. com/dingoes-do-bark-why-most-dingo-facts-you-think-you-know-are-wrong-68816.

Coyote bark alarm videos: George Bumann and Jenny Golding, "Play-list—Coyote Alarm Vocalizations," 2013, www.youtube.com/@ georgebumannjennygolding3663/playlists.

Alarm calls innate rather than learned: Eugene S. Morton, *Animal Talk: Science and the Voices of Nature* (New York: Random House, 1992).

CHAPTER 13: SPEAKING CHICKADEE-ESE

Vervet monkey study: R. M. Seyfarth, D. L. Cheney, and P. Marler, "Vervet Monkey Alarm Calls: Semantic Communication in a Free-Ranging Primate," *Animal Behaviour* 28, no. 4 (1980): 1070–1094, doi.org/10.1016/ s0003-3472(80)80097-2.

Animal dummies used in research: BBC, "Learn the Meerkat Language!" *BBC Earth* (2018), www.youtube.com/watch?v=NokF599u1Ck. M. Leroux, A. M. Schel, C. Wilke, et al., "Call Combinations and Compositional Processing in Wild Chimpanzees," *Nature Communications* 14, no. 2225 (2023), doi.org/10.1038/s41467-023-37816-y.

Chickadee alarms: C. N. Templeton, E. Greene, and K. Davis, "Allometry of Alarm Calls: Black-Capped Chickadees Encode Information About Predator Size," *Science* 308, no. 5730 (2005): 1934–1937, doi. org/10.1126/science.1108841. Jason R. Courter and G. Ritchison, "Alarm Calls of Tufted Titmice Convey Information About Predator Size and Threat," *Behavioral Ecology* 21, no. 5 (2010): 936–942, doi. org/10.1093/beheco/arq086.

Other birds interpreting chickadee alarms: C. N. Templeton and E. Greene, "Nuthatches Eavesdrop on Variations in Heterospecific Chickadee Mobbing Alarm Calls," *PNAS* 104, no. 13 (2007): 5479–5482, doi. org/10.1073/pnas.0605183104. N. V. Carlson, E. Greene, and C. N. Templeton, "Nuthatches Vary Their Alarm Calls Based Upon the Source of the Eavesdropped Signals," *Nature Communications* 11, no. 1, (2020): 526, doi.org/10.1038/s41467-020-14414-w.

Squirrels listening to bird alarms: M. V. Lilly, E. C. Lucore, and K. A. Tarvin, "Eavesdropping Grey Squirrels Infer Safety From Bird Chatter," *PLoS One* 14, no. 9 (2019): e0221279, doi.org/10.1371/journal.pone.0221279. K. A. Schmidt, E. Lee, R. S. Ostfeld, and K. Sieving, "Eastern Chipmunks Increase Their Perception of Predation Risk in Response to Titmouse Alarm Calls," *Behavioral Ecology* 19, no. 4 (2008): 759–763, doi. org/10.1093/beheco/arn034.

Why pishing doesn't work in tropics: Gary M. Langham, Thomas A. Contreras, and Kathryn E. Sieving, "Why Pishing Works: Titmouse (Paridae) Scolds Elicit a Generalized Response in Bird Communities," *Ecoscience* 13, no. 4 (2006): 485–496, doi.org/10.2980/1195-6860(2006)13[485:WPWTPS]2.0.CO;2.

Seet alarm: E. Natasha Vanderhoff and Perri K. Eason, "The Response of American Robins (*Turdus migratorius*) to Aerial Alarms," *Behaviour* 146, no. 3 (2009): 415–427, doi.org/10.1163/156853909X410982. N. T. Tegtman and R. D. Magrath, "Discriminating Between Similar Alarm Calls of Contrasting Function," *Philosophical Transactions of the Royal Society B* 375, no. 1802 (2009): 20190474, doi.org/10.1098/rstb.2019.0474.

Inability of raptors to hear *seet* calls: K. J. Jones and W. L. Hill, "Auditory Perception of Hawks and Owls for Passerine Alarm Calls," *Ethology* 107 (2001): 717–726, doi.org/10.1046/j.1439-0310.2001.00698.x.

Phone ringtones that adults can't hear: Paul Vitello, "A Ring Tone Meant to Fall on Deaf Ears," *New York Times*, June 12, 2006, www.nytimes.com/2006/06/12/technology/12ring.html.

Caterpillars imitating bird *seet* alarm: Jessica Lindsay, "Why Do Caterpillars Whistle? Acoustic Mimicry of Bird Alarm Calls in the *Amorpha juglandis* Caterpillar," ScholarWorks: University of Montana (2015): scholarworks.umt.edu/utpp/60.

CHAPTER 14: VARIATIONS ON A THEME

Prairie dog language: Con Slobodchikoff, *Chasing Dr. Dolittle: Learning the Language of Animals* (New York: St. Martin's Press, 2012). C. N. Slobodchikoff, A. Paseka, and J. L. Verdolin, "Prairie Dog Alarm Calls Encode Labels About Predator Colors," *Animal Cognition* 12, no. 3 (2009): 435–439, doi.org/10.1007/s10071-008-0203-y.

Exclusivity of language to humans: Marc D. Hauser, Noam Chomsky, and W. Tecumseh Fitch, "The Faculty of Language: What Is It, Who Has It, and How Did It Evolve?" *Science* 298, no. 5598 (2002): 1569–1579, doi.org/10.1126/science.298.5598.1569. Kentaro Abe and Dai Watanabe, "Songbirds Possess the Spontaneous Ability to Discriminate Syntactic Rules," *Nature Neuroscience* 14 (2011): 1067–1074, doi.org/10.1038/nn.2869.

Wild turkey communication: Joe Hutto, *Illumination in the Flatwoods: A Season Living Among the Wild Turkey* (Essex, CT: Lyons Press, 1998). Fred Kaufman, "My Life as a Turkey," PBS *Nature*, season 30, episode 4, released November 22, 2011 (based on Hutto, 1998).

CHAPTER 15: HOWLING MYSTERIES

Canines reading human signals: B. Hare and M. Tomasello, "Human-Like Social Skills in Dogs?" *Trends in Cognitive Sciences* 9, no. 9 (2005): 439–444, doi.org/10.1016/j.tics.2005.07.003. Z. Virányi, M. Gácsi, E. Kubinyi, et al., "Comprehension of Human Pointing Gestures in Young Human-Reared Wolves (*Canis lupus*) and Dogs (*Canis familiaris*)," *Animal Cognition* 11, no. 3 (2008): 373–387, doi.org/10.1007/s10071-007-0127-y.

Emission of light by the human body: M. Kobayashi, D. Kikuchi, and H. Okamura, "Imaging of Ultraweak Spontaneous Photon Emission From Human Body Displaying Diurnal Rhythm," *PLoS One* 4, no. 7 (2009): e6256, doi.org/10.1371/journal.pone.0006256.

Dogs detecting cremation ashes, cancer, and stress: Alissa Greenberg, "Dogs Sniff Out Cremation Ashes Amid Wildfire Destruction," NOVA *Next*/PBS, October 5, 2021, www.pbs.org/wgbh/nova/article/cremation-ashes-sniffing-dogs-california-wildfire. E. Cohen and J. Bonifield, "Meet the Dogs Who Can Sniff Out Cancer Better Than Some Lab Tests," CNN, February 3, 2016, www.cnn.com/2015/11/20/health/cancer-smelling-dogs. C. Wilson, K. Campbell, Z. Petzel, and C. Reeve, "Dogs Can Discriminate Between Human Baseline and Psychological Stress Condition Odours," *PLoS One* 17, no. 9 (2022): e0274143, doi.org/10.1371/journal.pone.0274143.

Dolphins detecting heart attack victim: Carl Safina, *Beyond Words: What Animals Think and Feel*, reprint edition (London: Picador, 2016).

Elephants knowing at a distance: Lawrence Anthony and Graham Spence, *The Elephant Whisperer* (New York: St. Martin's, 2009). D. Martin, "Lawrence Anthony, Baghdad Zoo Savior, Dies at 61," *New York Times*, March 11, 2012, www.nytimes.com/2012/03/12/world/africa/lawrence-anthony-baghdad-zoo-savior-dies-at-61.html. BBC, "A Herd of Elephants Marched 12 Hours to the House of Lawrence Anthony ..." Facebook, November 20, 2019, www.facebook.com/BBCOne/posts/a-herd-of-elephants-marched-12-hours-to-the-house-of-lawrence-anthony-after-he-d/2798498103503923.

CHAPTER 16: 1 + 1 = 1,000

Shapes of alarm: Jon Young, *What the Robin Knows: How Birds Reveal the Secrets of the Natural World* (Boston: Mariner Books, 2013).

Richard Louv referencing Jon Young: Richard Louv, *Last Child in the Woods* (London: Atlantic, 2010).

Impacts of aircraft on wild animals: Bernie Krause, *The Great Animal Orchestra*, reprint edition (New York: Back Bay, 2013).

CHAPTER 17: THE HONEYCOMB LANDSCAPE

Sitting for an extended period in nature: David G. Haskell, *The Forest Unseen*, reprint edition (London: Penguin Books, 2013).

CHAPTER 18: WILDSENSE 2.0

Clever Hans: Edward T. Heyn, "Berlin's Wonderful Horse; He Can Do Almost Everything but Talk—How He Was Taught," *New York Times*, September 4, 1904, www.nytimes.com/1904/09/04/archives/berlins-wonderful-horse-he-can-do-almost-everything-but-talk-how-he.html. Oskar Pfungst, *Clever Hans (The Horse of Mr. von Osten): A Contribution to Experimental Animal and Human Psychology*, trans. C. L. Rahn (New York: Henry Holt, 1911), www.gutenberg.org/files/33936/33936-h/33936-h.htm.

Deer responding to change in human scent: Geoffroy Delorme, *Deer Man: Seven Years of Living in the Wild*, trans. Shaun Whiteside (Vancouver: Greystone Books, 2022).

CHAPTER 19: BECOMING INVISIBLE

Going for a walk with a quiver full of arrows: Richard K. Nelson, *Hunters of the Northern Forest: Designs for Survival Among Alaskan Kutchin* (Chicago: University of Chicago Press, 1973).

Missing the gorilla: Daniel Simons, "Selective Attention Test," uploaded May 10, 2010, YouTube video, www.youtube.com/watch?v=vJG698U2Mvo.

Artist dates: Julia Cameron, *The Artist's Way* (New York: TarcherPerigee, 2016).

CHAPTER 20: THE POWER OF ONE

Jane Goodall and Dian Fossey: Jane Goodall, *In the Shadow of Man* (New York: Houghton Mifflin Harcourt, 1971). Dian Fossey, *Gorillas in the Mist* (New York: Houghton Mifflin Harcourt, 1983).

Baboons of Kenya: Robert M. Sapolsky, *A Primate's Memoir: A Neuroscientist's Unconventional Life Among the Baboons* (New York: Scribner, 2002).

Mule deer of Lander, Wyoming: Joe Hutto, *Touching the Wild: Living With the Mule Deer of Deadman Gulch* (New York: Skyhorse, 2014).

Book of American River natural history: Jo Smith, *The Outdoor World of the Sacramento Region* (Sacramento: American River Natural History Association, 1993).

Coyote attacks: Wikipedia, "Coyote Attack," last modified July 13, 2023, en.wikipedia.org/wiki/Coyote_attack.

Coyotes of American River, Sacramento: J. Junghans, "Coyote Guy," *Sacramento Magazine*, May 16, 2020.

Species loneliness and the age of loneliness: Richard Louv, *Our Wild Calling: How Connecting With Animals Can Transform Our Lives—and Save Theirs* (Chapel Hill: Algonquin Books, 2019). Krista Stevens, "How a Nonagenarian Insists We Can Avoid the Age of Loneliness," Longreads, April 21, 2020, longreads.com/2020/04/21/how-a-nonagenarian-insists-we-can-avoid-the-age-of-loneliness/.

CHAPTER 21: MAKING IT YOUR OWN

Stop moving: Tom Jay, *The Blossoms Are Ghosts at the Wedding* (Chimacum, WA: Empty Bowl Press, 2019).

Visiting Connie Hagar: Edwin Way Teale, *Wandering Through Winter: A Naturalist's 20,000-Mile Journey Through the North American Winter* (New York: Dodd, Mead & Co., 1965).

Fifty-nine million people feed birds in the US: Susan Morse, "To Feed or Not to Feed Wild Birds," US Fish and Wildlife Service, www.fws.gov/story/feed-or-not-feed-wild-birds.

Home waters concept: Todd Tanner, "Home Waters," *Hatch*, December 22, 2014, www.hatchmag.com/articles/home-waters/7712335.

Gilbert White: Gilbert White, *The Natural History of Selborne*, Anne Secord, ed. (London: Oxford World's Classics, 2013).

Tove Jansson: Tove Jansson, *The Summer Book*, trans. Thomas Teal (New York: New York Review Books Classics, 2008).

Robin Wall Kimmerer: Robin Wall Kimmerer, *Braiding Sweetgrass* (Minneapolis: Milkweed, 2015).

Numbered trees in Melbourne, Australia: Claire Dunn, *Rewilding the Urban Soul* (London: Scribe, 2021).

Rehabilitating urban habitat in Melbourne: J. Sparrow, "Undraining the Swamp: How Rewilders Have Reclaimed Golf Courses and Water-ways," *Guardian*, September 10, 2020, www.theguardian.com/ environment/2020/sep/11/undraining-the-swamp-how-rewilders-have-reclaimed-golf-courses-and-waterways.

Homegrown National Parks: Douglas W. Tallamy, *Nature's Best Hope: A New Approach to Conservation That Starts in Your Yard* (Portland, OR: Timber Press, 2020).

CONCLUSION

Tally of human languages: Stephen R. Anderson, "How Many Languages Are There in the World?" Linguistic Society of America, www.linguistic society.org/content/how-many-languages-are-there-world.